T0305623

Essentials of Mechatronics

Essentials of Mechatronics

John Billingsley
University of Southern Queensland
Queensland, Australia

A John Wiley & Sons, Inc., Publication

Published by John Wiley & Sons, Inc., Hoboken, New Jersey.
Published simultaneously in Canada.

Library of Congress Cataloging-in-Publication Data:

Billingsley, J. (John)
Essentials of mechatronics / by John Billingsley.
p. cm.
Includes bibliographical references and index.
ISBN-13 978-0-471-72341-7 (cloth)
ISBN-10 0-471-72341-X (cloth)
1. Mechatronics. I. Title.
TJ163.12.B55 2006
621–dc22

2005032762

Printed in the United States of America.

10 9 8 7 6 5 4 3 2 1

Contents

Preface ix

Acknowledgments xi

1. Introduction 1

1.1 A Personal View / 1
1.2 What Is and Is Not Mechatronics? / 6

2. The Bare Essentials 9

2.1 Actuators / 9
2.2 Sensors / 16
2.3 Sensors for Vision / 22
2.4 The Computer / 25
2.5 Interface Electronics for Output / 27
2.6 Interface Electronics for Input / 32
2.7 Pragmatic Control / 36
2.8 Robotics and Kinematics / 41

3. Gaining Experience 43

3.1 Coming to Grips with QBasic / 45
3.2 The Simplest Mobile Robot / 49
3.3 Ball and Beam / 56

3.4 "Professional" Position Control / 64
3.5 An Inverted Pendulum / 80

4. Introduction to the Next Level 91

4.1 The www.EssMech.com Website / 92

5. Electronic Design 95

5.1 The Rudiments of Circuit Theory / 95
5.2 The Operational Amplifier / 99
5.3 Filters for Sensors / 103
5.4 Logic and Latches / 113

6. Essential Control Theory 117

6.1 State Variables / 117
6.2 Simulation / 120
6.3 Solving the First-Order Equation / 121
6.4 Second-Order Problems / 123
6.5 Modeling Position Control / 125
6.6 Matrix State Equations / 127
6.7 Analog Simulation / 128
6.8 More Formal Computer Simulation / 130

7. Vectors, Matrices, and Tensors 131

7.1 Meet the Matrix / 131
7.2 More on Vectors / 132
7.3 Matrix Multiplication / 134
7.4 Transposition of Matrices / 135
7.5 The Unit Matrix / 136
7.6 Coordinate Transformations / 136
7.7 Matrices, Notation, and Computing / 138
7.8 Eigenvectors / 140

8. Mathematics for Control 143

8.1 Differential Equations / 143
8.2 The Laplace Transform / 146
8.3 Difference Equations / 150
8.4 The *z* Transform / 154
8.5 Convolution and Correlation / 157

9. Robotics, Dynamics, and Kinematics **161**

9.1 Gears, Motors, and Mechanisms / 161
9.2 Three-Dimensional Motion / 166
9.3 Kinematic Chains / 173
9.4 Robot Dynamics / 179
9.5 Simulating a Robot / 180

10. Further Control Theory **185**

10.1 Control Topology and Nonlinear Control / 185
10.2 Phase Plane Methods / 192
10.3 Optimization / 199

11. Computer Implementation **203**

11.1 Essentials of Computing / 203
11.2 Software Implications / 206
11.3 Embedded Processors / 210

12. Machine Vision **221**

12.1 Vision Sensors / 221
12.2 Acquiring an Image / 222
12.3 Analyzing an Image / 224

13. Case Studies **237**

13.1 Robocow—a Mobile Robot for Training Horses / 237
13.2 Vision Guidance for Tractors / 243
13.3 A Shape Recognition Example / 251

14. The Human Element **255**

14.1 The User Interface / 255
14.2 If All Else Fails, Read the Instructions / 259
14.3 It Just Takes Imagination / 260

Index **263**

Preface

There are many definitions of mechatronics, but most involve the concept of blending mechanisms, electronics, sensors, and control strategies into a design, knitted together with software.

With an abundant wealth of topics to choose from, authors of mechatronics textbooks are tempted to bundle them all into a massive compendium. This book seeks to throw out all but the essentials; although perhaps in hanging onto the baby, some bathwater will still remain.

There are a hundred ways of achieving all except the simplest of mechatronic design tasks. At every step, choice and compromise will be involved. Should a precision motor be used, or will a simple sensor and a sprinkle of feedback allow something cheaper and easier to do the trick? What does the end user ask for, really want, actually need—or eventually buy?

Specialists can handle the fine detail, the composition of the molded plastic, the choice of components for the electronic interface, machining drawings, embedded computer, or software development platform. At the top of the pyramid, however, there must be a mechatronic designer capable of making the design tradeoffs that will transform a client's demands or a bright idea into a successful commercial product.

In some ways, mechatronics is as much a philosophy as a science. At its heart is a way of looking at tasks that will, if necessary, achieve their objective by ducking aside into an alternative technology. The mechatronic engineer knows where to look for the side roads and has a shrewd idea of the merits of the diversion.

Acknowledgments

This book is the result of so many influences that there is a danger of this becoming the longest section. Perhaps I should start with the engineers of the autopilot industry who introduced me to the practical aspects of control system design. Laury Ambrose and Mike Skinner left me in no doubt as to their opinions of the quality of the servo loop designed with my new graduate academic skills.

Later, John Coales filled me with enthusiasm to research abstruse control methods such as fast-model predictive control. My team of Cambridge researchers, including David Hedgeland, John Moughton, Matthew Dixon, and Roger Kinns, led the charge to embed processor boards in the most unlikely applications.

In Portsmouth, life became even more exciting. Mechatronics and robotics abounded with the help of Harjit Singh, Fazel Naghdy, David Harrison, David Sanders, David Robinson, and many others. Arthur Collie lent the wisdom of years in industry to a passion for walking robots. Tim Dadd, now my son-in-law, joined me in meeting the problems of running a company that designed software for embedding in mass-produced appliances.

Australia has been fun. Sam Cubero, Jason Stone, Matt Petty, Stuart McCarthy, Brad Schultz, and others all pushed robotics forward, while Mark Phythian has taken up the cudgels of running Micromouse and Bilby contests. Mark Dunn has thrown himself into vision research, with more practical applications than you can shake a stick at.

The achievements and energy of my children Berry-Anne, Richard, and William have all helped to keep up my enthusiasm, while my wife Rosalind's play-writing successes have sometimes diverted my time to thespian activities.

1 Introduction

1.1 A PERSONAL VIEW

Although many writers are happy to put a date on the day a Japanese (or was it a Finn?) coined this rather ungainly word, *mechatronics* has been around in spirit for many decades.

My first brush with industry involved designing autopilots. The computers on which they were based used analog magnetic amplifiers—and later transistors—rather than the digital microcomputer we would expect today. Nevertheless, how can we describe as anything but a robot a machine that trundles through the sky, obeying commands computed from a multitude of sensor signals that enable it to make a perfect automatic landing?

By the mid-1960s, some computers had started to shrink. While the Atlas was fed a succession of jobs by an army of operators, an IBM1130, built into a desklike console, allowed real time interaction by the user. Soon we were able to buy "budget" single-board computers for a thousand British pounds. Although these had a mere 16 kilobytes (kbytes) of memory, their potential for mechatronics was immense.

One of my Cambridge researchers took on the task of revolutionizing the phototypesetter. The current state of the art was to spin a disk of letter images, triggering a flash to expose each letter onto photographic film. This was certainly "mechatronic" to an extent, requiring the precision positioning and timing under electronic control, but the new approach distilled the essence of mechatronics.

The method is now commonly found in the laser printer. A spinning mirror scans a laser beam across the photosensitive film, building up the image by rapid switching of the beam. Letter shapes are held in computer memory, and the entire mechanical design is simplified.

I consider this tradeoff between mechanics, electronics, and computing power to be the guiding principle of mechatronics.

The research team were soon knitting similar computers into a variety of real-time applications, including an "acoustic telescope" to build the signals from 14 microphones into an image of the source. Hydrofoils were simulated, violins were analyzed for their "Stradivarius-like qualities," and music was synthesized. A display for a color television, novel in those days, depended on a minimum of electronics and a wealth of software.

But computing power soon came in increasingly small packages. Texas Instruments had produced a single chip that could function as a pocket calculator. By the time I had moved from Cambridge to Portsmouth, Intel and Motorola were head-to-head with competing microprocessors.

In Britain, the Microprocessor Awareness Project (MAP) triggered a deluge of applications—but only a small proportion of them deserve truly to be considered as mechatronics.

Industrial firms were offered 2000 pounds'-worth of consultancy to consider how microprocessors could be added to their products. Some sharp operators made a killing, providing virtually identical reports to a diversity of clients. Others "brokered" projects to earnest academics. Printing machines sprouted boxes with twinkling LEDs (light-emitting diodes), wiring and relays patched on top of the "standard model." In many cases it made the machines virtually unusable and impossible to maintain.

Gradually, however, the concept percolated through that the computing aspect could be made fundamental to the operation of a machine. The mechanical precision and complexity could be traded off against electronics and computing power, just as in the case of the typesetter.

One MAP project concerned the design of a clock for a domestic cooker. Not very romantic, perhaps, but the client's choice of the primordial chip as used in the earliest pocket calculators made it a conundrum with attitude. It took several years and many generations of the product to persuade the company to adopt something simpler to program. The manufacturers of the original chip kept halving their price.

The chips were supplied, mask-programmed, in batches of 10,000. That concentrated the mind wonderfully on making sure that the code was correct. But once we had weaned the company off the TMS1000, there was room in the chip's memory not only for the job at hand but also for the next version we had in mind.

One focus of our research was the *Craftsman Robot*. An *energy regulator* is the switching element behind the knob that allows the power of a cooking ring to be varied. During its manufacture, several adjustments have to be made. We used a Unimation Puma 560 robot to pick each regulator from a tray and offer it to a test rig. Instead of acting as a simple "mover," however, the Puma was equipped with a screwdriver to adjust the regulator when it was still held in its gripper. Of course, we could not resist taking the robot apart and analyzing its software and drive circuitry.

Other industrial projects included marine autopilots and a flux-gate compass. But another interest would soon seize my attention.

In 1979, planning started for holding the Euromicro conference in London. Lionel Thompson, the chairman, wanted an added showpiece, and his mind was on "The Amazing Micromouse Maze Contest" that had just been announced by *IEEE Spectrum.* I put my hand up to organize the contest.

I then started to follow the news from the United States. Blows were nearly exchanged when the "dumb wall followers" sprinted through the maze from the entrance at one corner to the exit at the other, much faster than their brainier rivals. How could the rules be massaged to give brains the edge?

Donald Michie, a guru of technical conundrums, was all for making the objectives more abstract, perhaps adding a cat to the fray. The solution lay in the opposite direction, to give the mouse builders more specific information that could be designed into the logic of their machines. Our maze was specified as 16 × 16 squares, with the target at the center, not on the edge. In that way, paths could circle the center to form "moats" that no mere wall-follower could cross.

A preliminary run was held in Portsmouth in July, with results that literally gave me nightmares. Of the six mice that competed, only one could make any attempt to follow a passageway, let alone find the center. Japanese observers were there in force, cameras snapping away, and I was amazed that everyone seemed to enjoy the show.

At the conference in September, 15 mice competed. A sleek machine from Lausanne should perhaps have won—but it expected more precision of the maze than the carpenters had provided and became lodged on a join in the boards of the base.

The winner was a clanking contraption, cobbled together around a brilliant maze-solving algorithm that has remained relevant to this day.

The contest went from strength to strength, held in Paris, Tampere, Madrid, and Copenhagen, but for these first few years something struck me as strange. Not one of the winners was trained as an engineer. Great machines came from mathematicians, computer maintenance staff, and programmers for manufacturing industry, but engineers were notable by their absence.

In 1985 I was invited to Tsukuba, to see what the Japanese had made of the contest. There were 200 contestants, but the champion, Idani, was not an engineer in the formal sense. Later that year we took the contest back to the United States—the Japanese funded the trip to put some life back into an old adversary. A future champion was unearthed in MIT—but he was not then an academic; he was part of the laboratory staff.

So, what is it that defines a mechatronic engineer? What is the special aptitude that singled out these champions? What had they learned from their endeavors that was not to be found in a formal engineering course?

They were able to put together a concept in which strategy, computing hardware, sensors, electronics, and motors were blended together in harmony, not as a cobbled assembly of diverse technologies. Therefore we must distill the "good bits" from the diverse range of specializations that make up engineering as a whole.

Mobile robots are a fascinating application of mechatronics. A spinoff of the cooker clock project was the addition to our team of a seasoned researcher—a director of the company—who joined our Portsmouth research group to indulge his obsession with legged robots. Robug I rather ominously looked like a coffin on somewhat wobbly legs. Robug II shed all unnecessary weight and climbed walls. Together with Zig-Zag, it impressed the nuclear industry enough that they started placing orders for the design of robots for specific applications.

While we had been keen to give our robots intelligence, the last thing the clients wanted was for a robot, clambering on a nuclear pressure vessel with an angle grinder in its claw, to start showing initiative!

The market for these robots set a whole new direction for the company, newly emerged from the Tube Investments Group via a management buyout. Portsmouth Technology Consultants was born. I remained a director of the new company, even though by then I had moved to Queensland, Australia.

Ten years later, despite some major European funding for walking robot development, the company failed. The cloud had a silver lining. For scrap-metal prices, we were able to buy for the University of Southern Queensland the latest eight-legged walker, the result of a million dollars or more of development.

Although we had already developed an Australian ceiling walker all of our own, seen worldwide on BBC television, the research interest turned to agricultural applications, in particular to the vision guidance of tractors. With a videocamera, a computer, and a submodule for operating the hydraulic steering system, we were able to steer to an accuracy of better than an inch. The project was a technical tour de force, but a commercial failure. In hindsight, it is clear that the reason for the lack of sales was that we had set the price too low. Yes, too low.

We aimed to sell the system to dealers for $5000, for them to sell on at $10,000. That might appear to be a generous margin, but it was not enough. A purchaser might work a property many hundreds of miles from the dealer. A simple fault might render a quarter million dollar tractor unusable, and the dealer would be called out. After a lengthy journey, the dealer was still likely to be baffled.

A phoenix rose from the ashes of the project. An Australian company started to market a GPS (Global Positioning System) guidance system, one that displayed steering instructions to a human driver, at a price of many tens of thousands of dollars. A demand was swiftly seen for an interface between the GPS system and the actual steering of the tractor. The steering submodule that was a small part of the vision guidance system was just what was wanted. This time the price was set at several times the price of the entire original vision system, and sales were very good.

With a new commercial partner, we will soon combine vision with a low-cost precision GPS technique that we have developed. The project will be rolling again.

Another project with journalist appeal was Robocow—a nimble mobile robot for training horses for cutting contests.

In some ways, as technology advances the task of exploiting it becomes harder. The traditional approach to embedding some computing power was to take a microprocessor chip, add some supporting memory and interfaces, and then write the software "from the ground up." The concept of an "operating system" would be as alien as adding antilock braking to a rollerskate.

But when Webcams can be bought with drivers to interface them via DirectShow to Windows-based applications, how far up the evolutionary tree do you have to go to find your computing power? The price of a fully equipped PC card is today little more than that of an evaluation board for a Motorola HC12. Are we locked into complicated but popular technology "because it's there"? That is certainly the line we have been taking with a deluge of agricultural application opportunities. The data capture is quick and dirty, and we can concentrate on innovating ways to analyze it.

A project that appears strange—but actually makes good sense—is based on the ability to discriminate between animal species. When a sheep approaches a watering place, it is recognized and allowed to pass through a gate. When a feral pig comes the same way, it is also recognized and allowed to pass through an adjacent gateway, to another water source.

The difference is that the sheep will be allowed to go on its way after drinking, while the pig is confined until the farmer comes to pay it some serious attention. The economics of damage by feral pigs and the trade in feral pork are convincing reasons for funding the project.

The dynamic behavior of small marsupials is another area of interest. There is a breeding program for an endangered species of *sminthopsis*. The problem is that if the lady is not "in the mood," the animals are apt to kill each other. By tracking the movement of separated partners in adjoining cages, we hope to detect in real time when true love can take its course.

Texture analysis is usually a lengthy business, requiring substantial computing effort for correlations. Two applications require a speedy solution. The first is for the grading of oranges, where the extent of "goose bumps" on the surface is an indicator of quality.

The second is for the game of football. A speedy analysis of the status of the grass cover must be made, at least to avoid a lawsuit when an overvalued player slips on a bare patch and falls on his fundament. But is this really mechatronics?

So, what of the next generation of mechatronic engineers? How do we give them skill and ability with the essentials, without deluging them with the entire contents of the textbooks of at least three diverse disciplines? The Micromouse experience suggests that hands-on experimentation is an essential ingredient. While learning, software must be "crafted" by the student, rather than being ladled into the project as a bought-in commodity. The student must be prepared to deal with hydraulics or electromechanics, treating them as two sides of the same coin.

After the "bare essentials" whistle-stop tour of mechatronics, some experiments are presented that could whet the appetites of students to study the more detailed material that follows. "Seat of the pants" engineering will certainly get you started, but will go only so far.

Mechatronics is special. It is no more a mere mixture of electronics, mechanics, and computing than a *Chateau Latour* (or *Grange Hermitage*) vintage wine is a mixture of yeast and grape juice.

1.2 WHAT IS AND IS NOT MECHATRONICS?

Long ago, Caryl Capek wrote a book, *Rossum's Universal Robots.* It was as little about robotics as *Animal Farm* was about agriculture, but the term had been coined. Science fiction writers grew fat on the theme, and the idea of mechanical slave workers was lodged in the mind of the public.

When Devol designed a mechanical manipulator for Engelberger's firm, Unimation, it was endowed with the term "a robot arm." As a research topic, robotics ceased to be about tin men and turned to the articulation of mechanical joints to move a gripper or workpiece to a precise set of coordinates. The new "three laws of robotics" concerned the Denavit–Hartenberg transformation matrices, discrete-time control algorithms, and precision sensors.

Robotics is just a narrow subset of mechatronics. It is true that it has all the ingredients of sensing, actuation, and a quantity of computer-assisted strategy in between, but with every day the list of mechatronic products increases. In videorecorders, DVD players, jet airliners, fuel injection motor engines, advanced sewing machines, and Mars rovers, not to mention all the gadgetry that surrounds a computer, the jigsaw pieces of mechatronics are slotted together.

In something as simple as a thermostat, sensing and actuation of the heater are linked. But the element of computation is missing. It is not mechatronic. In automatic sliding doors, however, the criterion is not as cut and dried. A few simple logic circuits are enough to link the passive infrared sensor to the door motor, but the designer might have found that the alternative of embedding a microprocessor was in fact simpler to design and cheaper to construct.

Before 1960, autopilots were capable of automatic landing. Their computational processes were based on *magnetic amplifiers*, circuits using the saturation of a mumetal core with no semiconductor more complicated than a diode. As the aircraft approached its target, the mode switching from height-lock to ILS (instrument landing system) radiobeam to flareout controlled by a radar altimeter was performed by a clunking Ledex switch, a rotary solenoid driving something similar to an old radio waveband changer.

This must come close to qualifying as robotics, but lacking any trace of digital computation, it must fall short of mechatronics. For today's aircraft, however, with digital autopilots that can not only guide the aircraft across the world and land it, but also taxi it to the selected air bridge at the terminal, there can be no question that it is a mobile robot.

Machines that can roll, walk, climb, and fly under their own automatic control have come to share the title of robots, mobile robots. One example of such a robot is the Micromouse, which will be mentioned several more times in this book. *IEEE Spectrum Magazine* and David Christiansen must take the credit for devising a contest in which small trolleys explore a maze. I would like to claim personal credit for redefining the maze design and rules to give victory to the "intelligent" mouse, rather than the "dumb wall followers."

Many early Mice used stepper motors to move and steer them, controlled by microprocessors of one sort or another. The maze walls were sensed by a variety of photoelectric devices, although in at least two cases mechanical "feelers" were used with great success. To navigate through the maze, a map had to be built up in the microcomputer's memory. To solve the maze, a strategy was required. A further aspect of the software was the need to apply control to keep the mouse straight as it ran through the passageways. So, in one not-so-simple contest, all the ingredients of mechatronics were brought together.

The contest runs regularly to this day. Many of the early champions are still at the forefront, while simplified versions of the contest have been developed to encourage young entrants. While the experts hone their expertise, however, the bar has to be set lower and lower for the newcomers. Simply running through a twisted path with no junctions is a testing problem for most schools' entrants.

So, what is the "mechatronic approach"? How would a mechatronics engineer design a set of digital bathroom scales? Would they be based on a strain-gauge sensor, on the "twang" frequency of a wire tensioned by the user's weight, or on some more subtle piece of ingenuity?

When I opened up the machine on my bathroom floor, I was disappointed to discover that the pointer of a conventional mechanical scale had simply been replaced with a disk with a notched edge. As it rotated under the weight of the user, an *incremental optical encoder* counted the notches of the disk as they went by and displayed the count on a luminous display.

For a manufacturing company with an established market in mechanical scales, the "pasted on" digital feature makes sense. However a "truly mechatronic" solution would find a tradeoff between digits and mechanical precision that would simplify the product.

A hairdryer marketed some years ago featured a "bonnet," coupled by a hose to the hot-air unit. A plastic knob could be rotated to give continuously variable temperature control. So, how would you go about designing it? When the question is put to university classes, it always brings answers featuring potentiometers, thyristor power controllers, and often a microcomputer.

The product was actually much simpler. The airflow was divided into two paths after the fan. In one path was a heating element, regulated by a simple thermostat just "downstream," while the other simply blew cold air. The ornate knob moved a shutter that closed off one or other flow, or allowed a variable mixture of the two.

Good design can often demand an awareness of how to avoid excessive technology.

2

The Bare Essentials

2.1 ACTUATORS

A mechatronic system must "do" something, even if it is just to move a pointer or change a display. The industrial robot must have motors with which to move an end effector, perhaps a gripper, while another system's output might concern heaters.

The mechatronic engineer should not be in too much of a hurry to run to the catalog to choose an electric motor. To the electrical engineer, motors are a fascinating playground around which to debate the merits and challenges of axial flux, windage losses, rotor resistance, or commutation. The mechatronic engineer is by no means certain that the solution does not instead lie with something hydraulic or pneumatic.

This section attempts to put a selection of the vast range of actuators into some sort of perspective.

2.1.1 Choosing a Technology

The first question to ask is: "What must the output do?"

At the bottom end of the list, in terms of power, is the task of displaying a value on an indicator. Many automobile instrument panels have now been taken over by liquid crystal displays, probably putting them outside the grasp of mechatronics, but they are just the tip of the iceberg.

For many years the simplest of cheap automobile instruments, such as the fuel gauge, have been moved by a bimetal strip. Around it is wound some resistance wire. As current is passed through the wire, the temperature rises and the bimetal bends to move the pointer. A simple twist compensates for variation in ambient temperature. This old technology has been given a new lease on life by the arrival of *memory alloys* that change their shape with temperature.

For many applications, this simplicity and robustness is ruled out of question by a need for a rapid response. An electromagnetic solution might have more appeal.

When current flows in a conductor within a magnetic field, the conductor experiences a force. That more or less sums up electric motors! But the devil is in the detail.

In an electromagnetic indicator, the force is opposed by a spring, so that the deflection of the needle increases with the current.

The simplest electromagnetic actuator that can move a load is the solenoid. When current passes through the solenoid's winding, it results in a magnetic field that causes a slug of soft iron to move to close a gap in the magnetic circuit. This single action might be enough, say, to release a remote-entry door lock. But other applications demand something more versatile.

2.1.2 DC Motors

You are probably most familiar with the permanent-magnet DC motor, used in everything from toys to tape recorders. The rotor is wound in such a way that the electromagnetic force causes the rotor to rotate. If the currents in the motor's conductors were constant, the rotor would move to some stable position, swing to and fro around it a few times, and then come to rest. But the current is not allowed to be constant. Long before the stable position is reached, a commutator breaks the current to that particular coil and energizes the next one in succession. The motor continues to rotate.

The "old-fashioned" structure of the commutator used curved plates of copper with brushes, often made of carbon, that rubbed on them. The "brushless" DC motor is becoming increasingly common. Here sensors measure the rotor position, and electronic switches apply the commutation by selecting the appropriate coils. There is another important difference. Since the magnetic material usually has more mass than the rotor, in a traditional motor it is the coils that rotate. In a brushless motor the coils are fixed and the magnet rotates.

2.1.3 Stepper Motors

Now let us take away the commutation again. Energize one coil and the rotor is pulled to a particular position, requiring a fair amount of torque to deflect it. Energize another coil and the motor "steps" to another position. In other words, by selecting coils in sequence, a computer can step the motor an exact number of increments to a new position—this is a "stepper motor."

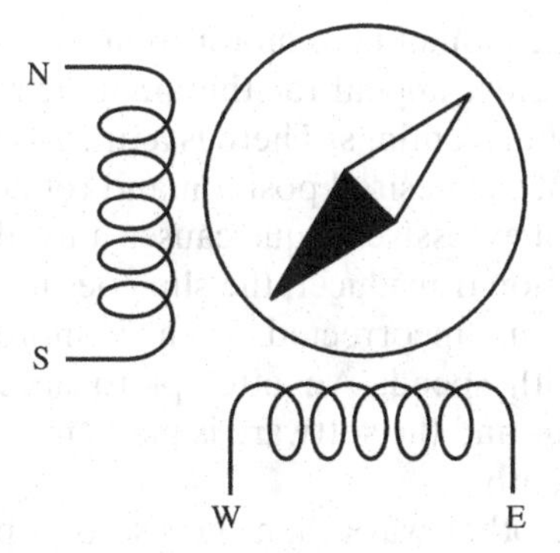

Figure 2.1 Stepper schematic—NSEW.

Think of it in terms of a compass needle being pulled into line by a pair of coils, arranged north–south and east–west (see Fig. 2.1). Current can be passed through these coils in either direction, so we might start with both the NS and EW coils being driven in the "positive" direction, resulting in the needle pointing northeast. Now if we reverse the drive the NS coil, the needle will move to point southeast. Reverse the EW coil, and it will rotate to point southwest. Make the drive to the NS winding positive again, and the needle moves on to point northwest. Finally reverse the EW drive to be positive and the needle completes the circle to point northeast once again. You can see an animation of this at www.essmech.com/2/1/3.htm

In practice, the magnet of a stepper motor has a large number of poles, and the windings are helped by a similar large number of salient polepieces (Fig. 2.2) in the soft iron on which they are wound. As a result, the switching sequence must be repeated 50 times for a "200-step" motor to make one complete revolution.

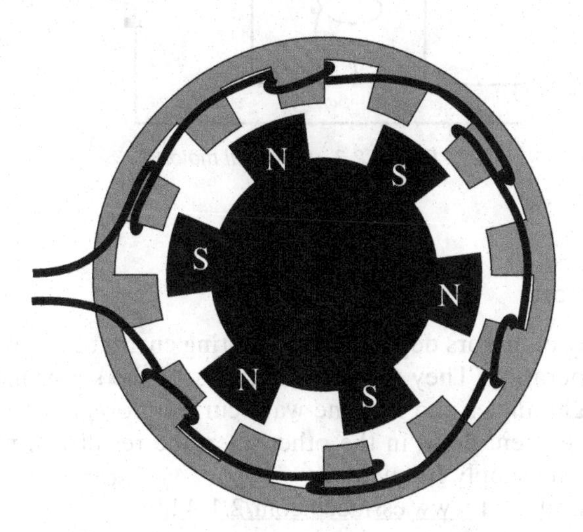

Figure 2.2 Stepper schematic—polepieces.

Simple software can command the motor to move to a desired position, so the stepper motor has great appeal for the amateur robotics builder. But it has a great number of shortcomings. There is a limit to the torque it can resist before it "clunks" out of the desired position and rotates to a different stable location. If a transient of excessive torque causes it to "drop out of step", then, without a separate position transducer, the slip goes unnoticed by the processor and the error remains uncorrected. What is more, this dropout torque decreases markedly with speed. An attempt to accelerate the motor too rapidly can be disastrous and the software is made more complex by the need to ramp the speedup gently.

Of course, there are other ways than the use of a permanent magnet for producing a magnetic field. More powerful DC motors, such as automobile starter motors, use current in a *field winding* to generate the stator's magnetic field. Similar motors are not restricted to using direct current. By connecting the stator and rotor windings in series, the torque will be in the same sense whether positive or negative voltage is applied across it. The motor can be driven by either an AC or DC voltage. This is the *universal motor* (Fig. 2.3), to be found in vacuum cleaners and a host of other domestic gadgets.

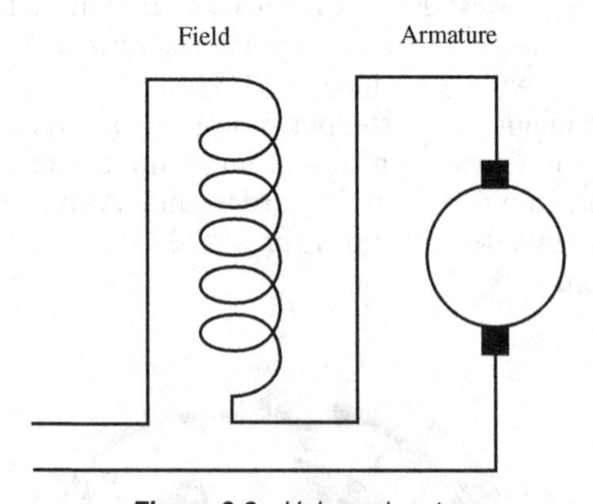

Figure 2.3 *Universal motor.*

2.1.4 AC Motors

Another family of motors depend on alternating current for their fundamental mode of operation. They use *rotating fields.* If the stator has two sets of windings at right angles and if a sine-wave current flows in one winding and a cosine-wave current flows in the other, then the result is a magnetic field that rotates at the supply frequency.

This is illustrated at www.essmech.com/2/1/4.htm.

From this one simple principle, a host of variations are possible. In one case, short-circuited coils are wound onto a soft-iron rotor. If the rotor is stationary, the rotating field induces currents in the rotor coils that in turn propel the rotor to rotate with the field. So the rotor accelerates, but cannot quite catch up with the field. If the rotor were to rotate at the supply frequency, it would experience no relative rate of change of field and no current would be induced in it.

The rotor "sees" the *slip frequency*, the amount by which the rotation falls short of the field rotation. Large industrial motors are designed to give maximum torque for a few percent of slip, thus improving their efficiency but requiring some special provision to get them up to speed (see Fig. 2.4).

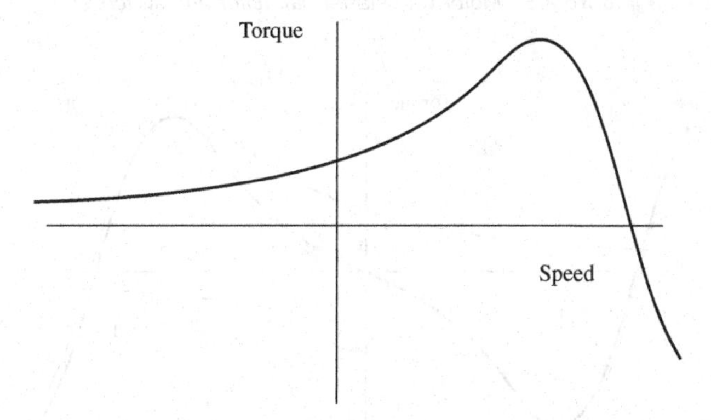

Figure 2.4 *Torque–slip curve.*

Induction motors can make useful servomotors. If the sine-wave winding is powered "at full strength" while the cosine-wave current is of a variable magnitude, then the rotating component of the field can be varied in strength, including the possibility of reversing its direction. This variable field servomotor has sufficient torque to move aircraft control surfaces. Now, however, the torque–slip characteristic must be modified so that maximum torque is generated at 100% slip—when the motor is at a standstill.

It is possible to run a two-phase induction motor from a single-phase supply. One phase is connected across the supply, while the second is energized via a series capacitor. This capacitor gives the phase shift that is needed to result in a rotating field. However, many appliances such as water pumps have a switch to disconnect the capacitor as soon as the motor is "up to speed," so that the motor continues running from a single phase alone. It works as shown in Figure 2.5.

The field from a single phase can be thought of as the result of two fields rotating in opposite directions (see curves in Fig. 2.6). The motor has a torque–slip characteristic that gives maximum torque for a small slip. Thus the torque from the field going in the "correct direction" is much greater than that of the opposite direction, so that the motor continues to rotate efficiently.

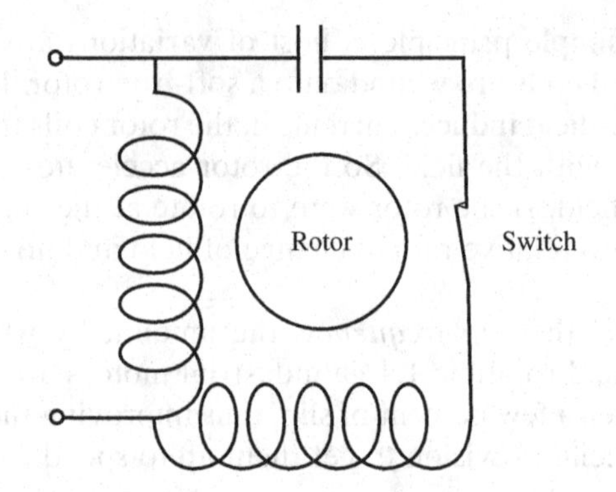

Figure 2.5 Motor with starter capacitor and switch.

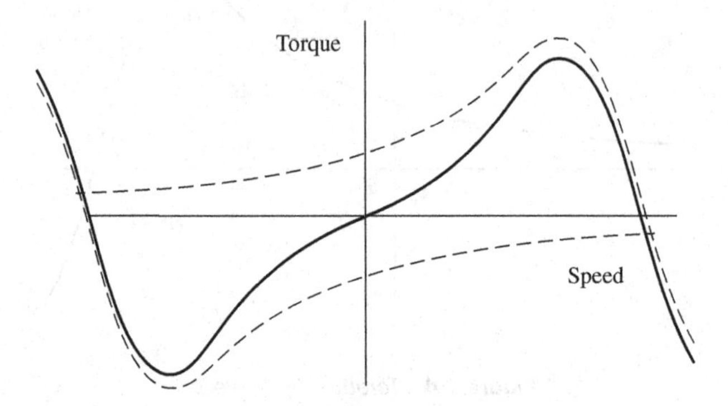

Figure 2.6 Combination of two torque–slip curves.

If a servomotor were designed with an "efficient" torque–slip curve, it would be in danger of running away, failing to stop when the control voltage was removed. But the conventional two-phase induction motor has some simple uses, as you will see in the section on interfacing.

Large induction motors are wound for three-phase operation, and their interfacing for control applications, such as in an elevator, presents problems all of their own.

The rotor of another variation of motor contains no iron and consists merely of a thin cylinder of copper. It will still experience rotational forces after the fashion of an induction motor. This is the "drag cup" motor, popular in servo repeater systems as used in autopilots in the 1960s. Although its torque was small, its tiny moment of inertia meant that response speeds could be very rapid.

Soft iron acquires a magnetic moment when in the presence of another magnetic field, but loses it when the field is removed. A permanent magnet is made of a "hard" material in which the magnetic moment can be self-supporting. Somewhere between these extremes is the material used in a *hysteresis motor.* The rotating field induces a magnetic moment in the rotor that remains, even as the rotating field advances. The field thus drags the rotor after it. Even when the rotor has accelerated to rotate in synchronism with the field, the residual permanent magnetism will keep the hysteresis motor rotating.

We can double the number of poles on the stator, so that the motor's basic speed of rotation is halved. The variations are endless. A motor can be constructed without bearings as a pair of rings, to be mounted on opposite sections of a robot joint. The structure can involve fields that are radial, as in a "conventional" motor, or fields that run parallel to the axis of rotation.

2.1.5 Unusual Motors

If you think that axial field motors are rare, rescue a $5\frac{1}{4}$-in. floppy disk drive from the junk heap. There is an easily recognized stepper motor for moving the head in and out, but where is the motor that rotates the floppy?

When you remove a plate from the large circuit board, you will see copper windings "stitched" to the board like the petals of a flower (see Fig. 2.7). The plate that covered it was in fact the magnetic rotor that carries the floppy round with it, while sensors on the board switch the fields to control the rotation. This axial field motor is truly "embedded" in the product, rather than being added as an identifiable component.

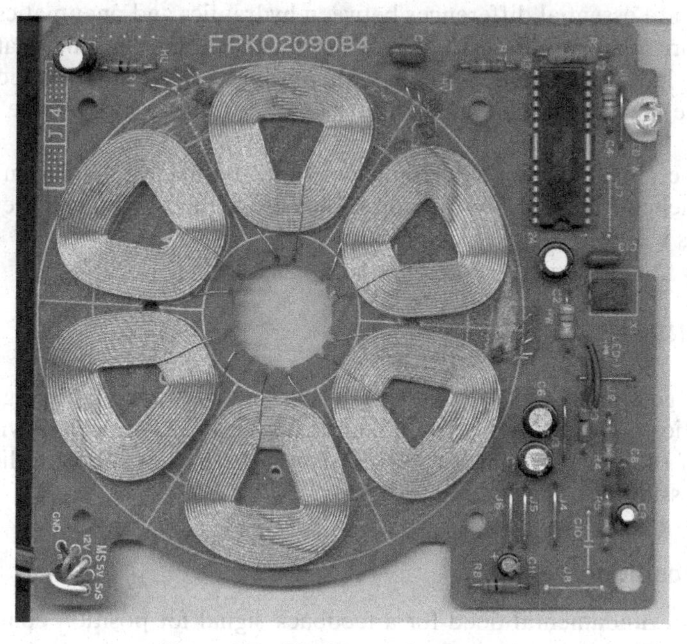

Figure 2.7 *Windings on floppy board.*

A motor can even be "rolled out flat." A linear stepper motor has its pole-pieces side-by-side. It is mounted close to a linear track stacked up from "slices" of magnetic and nonmagnetic material, along which it steps its way at considerable speed.

A linear induction motor can be propelled along a conducting or magnetic plate. This is a popular form of propulsion for hovering or magnetic levitation trains.

So, the mechatronic designer has much more to worry about than finding a motor in a catalog. Why should the motors be electrical at all? How about hydraulics and pneumatics?

2.1.6 Hydraulics and Pneumatics

The fundamental principles of these seem to be glaringly obvious. First, you must construct a cylinder and place a piston in it, maybe resulting in something not very different from a bicycle pump (see Fig. 2.8). When you pressurize the air or oil in one end of the cylinder, the piston will be forced away. Once again, the details make a simple situation very complicated.

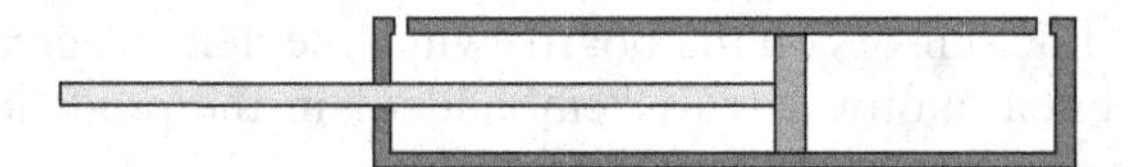

Figure 2.8 *Hydraulic/pneumatic cylinder.*

There are essential differences between hydraulics and pneumatics. Air is much more compressible than oil, but has much less inertia. Pneumatics will therefore have the edge in situations where rapid acceleration is needed, but where the power is not large. Hydraulics will flex its muscles for the heavier tasks.

But the choice of motor technology cannot be made in isolation. Power and efficiency will be just one factor, while ease of interfacing and the control dynamics will require just as much attention.

2.2 SENSORS

If the range of actuators seemed vast, it does not compare with the gamut of possibilities offered by sensors. Of course, the field is narrowed down by the nature of the quantity that is to be measured. Perhaps it is best to list some of the possibilities.

2.2.1 Position

There is a fundamental need for a feedback signal for position control, but the choices are numerous.

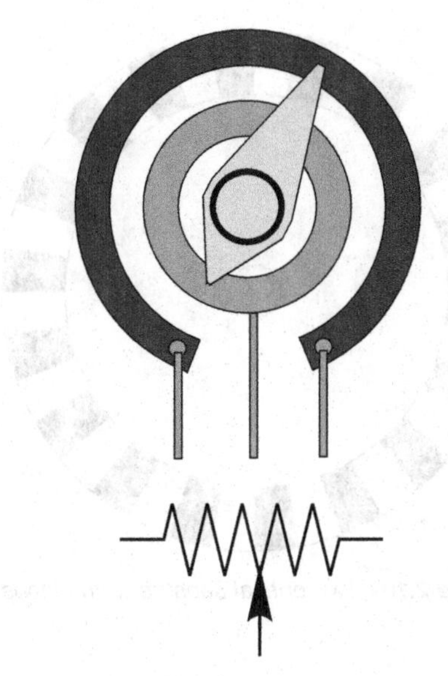

Figure 2.9 Potentiometer.

The "classic" sensor is the potentiometer (Fig. 2.9), where a voltage is applied across a resistive track and a moving "wiper" picks off a proportion of the voltage corresponding to its position. The track can be made from a winding of resistance wire, or a track of conducting plastic or carbon composition. The motion sensed can be linear or rotary. Prices vary by factors of hundreds, influenced by noise, reliability, and required lifetime.

The plastic potentiometer gives a signal that is not quantized—it varies smoothly and not in steps. This analog property has many advantages when the objective is simplicity, but like all analog signals, there is a question of accuracy and linearity.

Other sensors are "steppy" by nature, where the steps are absolutely defined by the construction of the sensor. One such sensor is the *incremental encoder.* Yet again, several technologies offer themselves, but the one most frequently found is optical.

Think of a sequence of "stripes" (Fig. 2.10) moving between a light source and a phototransistor. The associated electronics can "see" the difference between stripe and gap, and can count the stripes as they move past. This count is presented as the measured position. Now, this is all very well if we are certain of the direction of movement, but clockwise or anticlockwise (counterclockwise) movement will give signals that are indistinguishable from each other. So, how can they be resolved to count up or down?

The answer is to provide a second phototransistor alongside the first to render the transducer *two-phase* (see Fig. 2.11). The signals change one after the other. Now if we give the values 0 or 1 to each of the pair of signals, we might see the sequence of values 00, 10, 11, 01, 00, and so on for the pair when

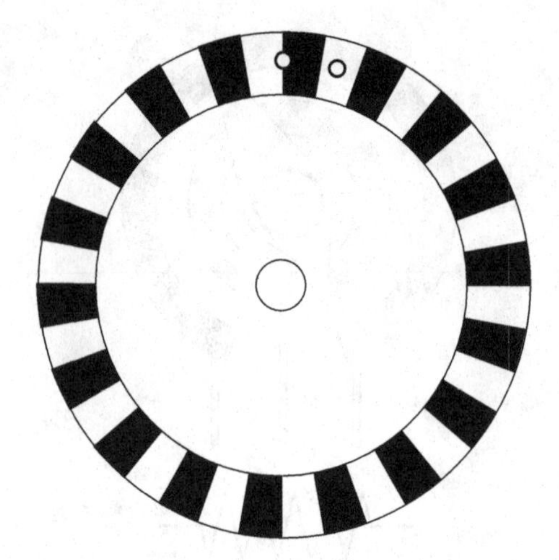

Figure 2.10 *Two optical sensors, with stripes.*

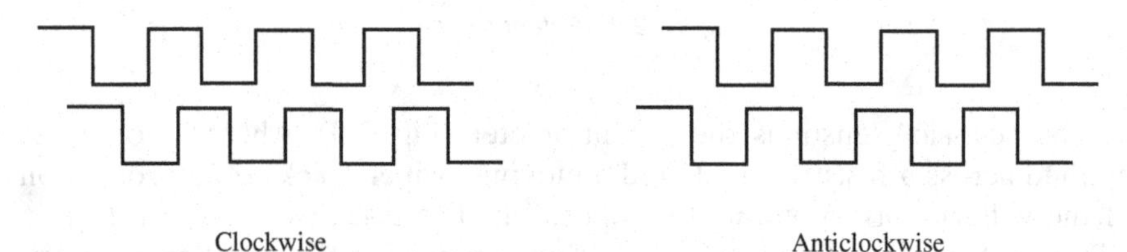

Figure 2.11 *Waveforms traveling forward and reverse.*

the rotation is clockwise. If it is rotated anticlockwise, however, it would be 00, 01, 11, 10, 00, and so on. Look closely, and you will see an essential difference, one that will cause a logic circuit or a couple of lines of software no trouble at all.

Even within optical sensors, there are many choices. The resolution can be increased to many thousands of "stripes" per revolution by passing two optical gratings across each other and observing the change in the Moiré pattern. An even finer resolution can result from mixing laser signals to produce interference fringes.

Now, this incremental technique seems an ideal solution—until you realize that it tells you only the change in position, not the absolute position at switch-on time. To find the initial position, one solution is to run the system to some hard stop and reset the counter there.

A more subtle technique was used in the Unimation Puma robot. As well as the incremental stripes, of which there were around a hundred per revolution of the geared motor, an extra stripe was added so that one specific point in the revolution could be identified. At startup, the motor rotated slowly until this stripe was detected. This still left tens of possibilities for the arm position,

Figure 2.12 Rotary Gray code.

as the motor was geared to rotate many times to move the arm from one extreme to the other. But in addition to the incremental transducer, the arm carried a potentiometer. Although the potentiometer might not have the accuracy required for providing the precision demanded of the robot, it was certainly good enough to discriminate between whole revolutions.

The concept of "coarse–fine" measurement is a philosophy in itself!

Of course, there is the alternative possibility of reading 10 binary signals at once, giving a 10-bit number with an "at a glance" resolution of 0.1%. However, the 10 tracks to give these signals must be very accurately aligned, even when the output is arranged as in *rotary Gray code* (Fig. 2.12). In Gray code, the first stripe is a half-revolution sector. The next is also a half-revolution sector, displaced so that the two outputs divides the range into four quadrants. Then each successive track has double the number of stripes, with edges splitting in half the regions defined so far. The last has 512 fine stripes. These devices are not cheap!

Other sensors can be based on magnetic "Hall effect" semiconductors of the switching variety for locating an endstop or for counting teeth in a magnetic gear or track.

Alternatively, analog Hall effect sensors (see example in Fig. 2.13) can pick off the sine and cosine of a rotation angle, by detecting the component of a magnetic field perpendicular to their sensing plane. These are "noncontact" and have high reliability.

A chip is available from Phillips, the KMZ43T (the superseded version was the KMZ41), which gives sine and cosine outputs depending on the angle of the field. Provided it is strong enough, they are independent of its strength. The output gives two electrical cycles per revolution of the magnet.

A noncontact sensor that perhaps does not entirely deserve its popularity is the *linear variable differential transformer* (LVDT) (Fig. 2.14). A coil is

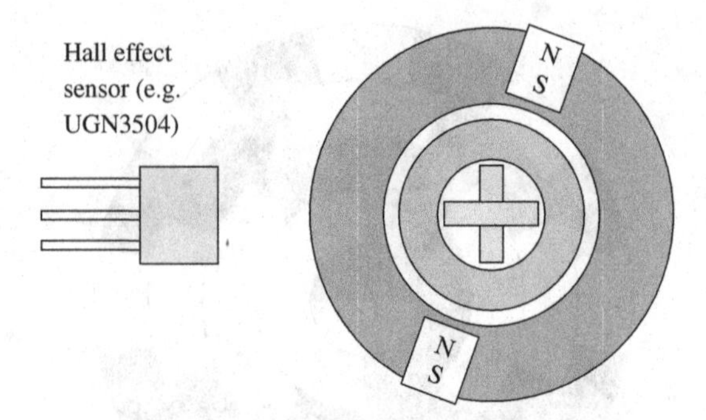

Figure 2.13 *"Crossed" Hall effect angle sensor.*

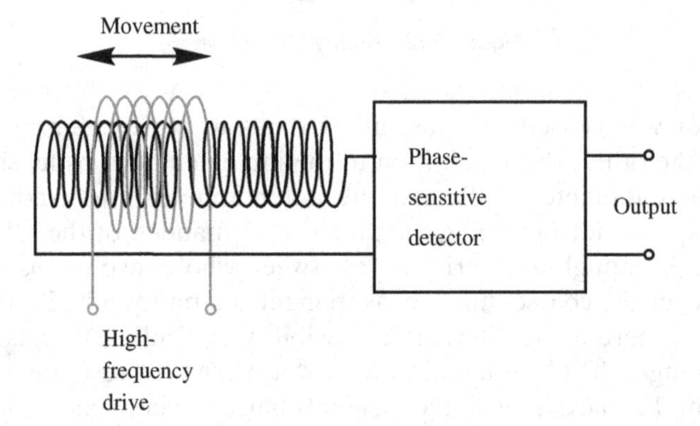

Figure 2.14 *LVDT.*

energized with a high-frequency signal, while another coil wound over it and sliding along it picks off a varying voltage by induction. The circuitry includes oscillator and detector, and the result is an output voltage that is proportional to the displacement.

There is a common displacement transducer that gives a resolution of a fraction of a millimeter in two dimensions, yet costs only $20 or less. It is used in the computer "mouse" and is based on the optical transducer described above. An alternative "optical" mouse has a simple image chip that "looks" at marks on the desk underneath the mouse and works out the movement by correlation.

If you widen your scope, the examples keep coming. GPS receivers must be included. By skillful combination of the carrier-phase measurements from two receivers, displacements can be tracked to subcentimeter accuracy. Distances can be measured with radar, with ultrasonic "estate-agent-grade" room measuring instruments or with the Polaroid sensors used in early autofocus cameras. The time of flight of laser pulses is used in another family of distance meters and in the Sick sensor.

2.2.2 Velocity

The response of most position controllers can be made more "crisp" by the addition of a velocity signal in the feedback. Now this can of course be derived by computation from a position sensor, although this "secondhand" signal is more sluggish than a direct measurement.

Any small permanent-magnet DC motor will act as a generator. When coupled to the shaft of a servomotor, it will give a voltage proportional to the velocity of rotation. The voltage will be somewhat "lumpy" as a result of the commutation, but will still be a great help to stability. The result is known as a "tacho" (*tachometer*) signal.

In an agricultural tractor, it is common to measure speed over the ground by radar. The unit is mounted below the body and detects the Doppler shift of a radar pulse directed obliquely at the ground.

Another form of velocity can be important for navigation: angular velocity. There are many forms of rate gyroscope to measure it. In classic form, these contain an inertial mass that spins at high speed. When there is a rotation about an axis perpendicular to that of the spin, it will result in substantial precession forces that are easy to measure. These rate gyroscopes have been an essential component of autopilots from early times.

More recent devices use a miniature "tuning fork" and have no rotating parts. As the fork rotates about its long axis, the to-and-fro vibration of the tines will gain a side-to-side component that can be measured. A similar device uses vibrations in a ring.

Yet another more recent device is based on the velocity of light in a coil of optical fiber.

2.2.3 Acceleration

A transducer for linear acceleration can take the form of a tilt sensor, measuring the displacement of a pendulum or of a mass mounted on leaf springs. This is really a secondhand measurement, where the acceleration is deduced from a position measurement of the movement of the pendulum. However, it is different from inferring the acceleration from measuring the position of the accelerating vehicle and computing differences.

Other sensors are designed to measure high-frequency vibrations. One range of commercial sensors is a close relative of a microphone, using either the change in capacitance between two moving plates or a piezoelectric voltage as a small mass resists the disturbing acceleration.

In a similar way, rotational acceleration can be deduced from the torque required to rotate a miniature "dumbbell," or from two linear accelerometers positioned some distance apart.

2.3 SENSORS FOR VISION

Optical sensing systems are many and various, with a variety of resolutions in both position and intensity. In order of increasing complexity, they can be ranked as follows.

2.3.1 Single Point, Binary

At the bottom of the ladder is a single *phototransistor*. When light falls on a phototransistor, it allows current to flow through it to cause a change in the output voltage.

A "pair" consisting of a single light-emitting diode (LED) and a single phototransistor can be used in a variety of ways. As a *reflective opto switch* they can detect a dark mark on a light background or vice versa. To obtain any kind of picture the sensor would have to be scanned mechanically in both directions.

As a *slotted opto switch*, the LED and the phototransistor are mounted to face each other and indicate when there is an obstruction in the slot. There are no real vision applications of this configuration, but it makes a useful detector for motion measurement (see example configuration in Fig. 2.15). Two such sensors monitor the movement in each axis of a conventional computer mouse.

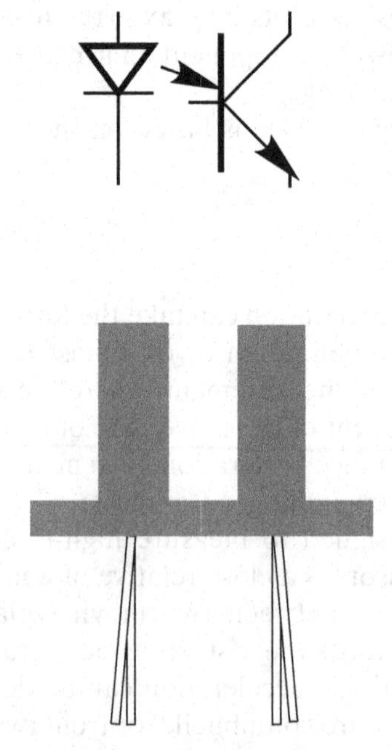

Figure 2.15 *Slotted opto switch.*

A simple "beambreak sensor" can be very versatile. It can calibrate the motion of a toolpiece moved by a robot, check the length of a drill, or verify that a hole has been punched.

2.3.2 Linescan Devices

A linescan sensor is used in every fax (facsimile) machine. The image of the page is focused onto the device, which "sees" a single scanline across the page. As the page is fed through the machine, the image is built up, line by line. This sensor could have around 2000 *pixels*, individual sensing elements, and will probably use the "bucket brigade" principle.

When light falls on the sensitive silicon sensor, a current flows that is proportional to the brightness. In the single phototransistor sensor the current is measured directly. In most camera systems, the current builds up a charge on a small embedded capacitor—part of the device. This allows the device to be much more sensitive, provided we are prepared to allow a little time to pass so that the charge can build up before we read and reset the voltage.

In a *bucket-brigade* device, we must apply a pulse sequence that shunts all the charges simultaneously into a second row of capacitors. Then a second sequence of pulses causes the charges to march along the second line of capacitors—living up to the title "bucket brigade." At the end of the chain, a sequence of voltages is seen, representing the light integrated by cell1, cell2, cell3, . . . and so on in turn. This kind of sensor needs an assortment of clock pulses to be accurately controlled in both timing and voltage. Fortunately, there are some other devices that are much easier to use.

The TSL214 and its successors use the same principle, in that the photoelectric current of each of their sensors is integrated in a capacitor. However, the method of reading the charge is much more robust and tolerant of timing and waveform variations. Each time a clock waveform rises and falls, a single "1" is moved along a shift-register. As the "1" reaches each cell, its capacitor charge is transferred to the output amplifier, so that a corresponding voltage appears on the output pin. One more input is needed, to insert the single "1" at the beginning of the sequence.

The chip runs from a single 5-V supply, and only two output bits from a microcomputer are needed to control it. It can easily be attached to the printer port of a PC, and the pulses can be generated by simple software. That leaves only the problem of inputting the brightness data to the PC.

An obvious option is to use an analog-to-digital converter (ADC)—perhaps an 8-bit device that can discriminate 256 levels of brightness. In many applications, though, the brightness needs to be compared only against a fixed level.

It is much easier in this case to perform the analog comparison in a single comparator and input a single bit per pixel. The method is a little less crude than it appears, since the sensitivity can be varied from within the software. By allowing more time to elapse between scans, a higher voltage will be obtained for a given brightness.

But we are already allowing this brief introduction to be complicated by considerations of interfacing.

2.3.3 Framescan Devices

These devices are built around a videocamera similar in principle to that of a camcorder. Some texts still refer to the vacuum-tube cameras that are now part of TV history. The "useful" technology is at present centered on a photosensitive silicon chip.

An array of photosensitive sites on the chip allow charge to build up on capacitors (also part of the chip) at a rate proportional to the light falling on each pixel, just as in the case of the linescan camera. By a variety of bucket-brigade operations, these charges are sampled in turn and appear at an output pin essentially in the form of a raster TV signal.

There is another, simpler form of camera termed a "Webcam." It usually has fewer pixels than does a videorecorder camera, typically 640 × 480. Its virtues are that it is cheap and that it comes with a complete interface to a personal computer.

A color TV display has a fine array of red, green, and blue phosphor dots, which have the combined effect of creating a full-color image. More expensive cameras use three chips, one to receive each primary color. The cheaper versions, including Webcams, use a single chip.

By printing a tinted pattern on the face of the camera chip, the manufacturers can produce an output that can be converted into a full-color picture, in exchange for some loss of resolution. Groups of 4 pixels are arranged with 2 of green and 1 each of red and blue (e.g., see Fig. 2.16). Since the eye resolves color with much less sharpness than it does brightness, the lower resolution of the color data is not a problem.

```
R G R G R G R G R G R G
G B G B G B G B G B G B
R G R G R G R G R G R G
G B G B G B G B G B G B
R G R G R G R G R G R G
G B G B G B G B G B G B
R G R G R G R G R G R G
G B G B G B G B G B G B
```

Figure 2.16 *Red-green-green-blue (RGGB) array.*

2.4 THE COMPUTER

Computers come in all shapes and sizes. Some are designed to be embedded in mechatronic products, while others are virtually complete products in themselves.

But the sophisticated PC and the simplest microprocessors have the same principles at heart. The essential components are memory and a *processor* consisting of an *arithmetic–logic unit* and a control unit. Bytes are "fetched" from the memory and treated either as data, numbers to be crunched, or instructions to be executed. The "cunning" of the computer (let us not yet call it "intelligence") lies in the ability to choose a different sequence order of instructions to obey, according to the value of the data.

So at rock bottom the program consists of a list of bytes to be executed. Switch on your Windows PC, click on START and RUN, and type "debug." Then, when a black panel appears, type

```
d f000:0
```

followed by clicking on <enter>

An array of two-digit numbers will appear, where the digits include the letters A–F. If the numbers are all zero, try inputting a different number to look at a different part of memory. When you have found something that looks interesting, try

```
u f000:0
```

or whatever number you have chosen.

Now a cryptic list of codes will appear on the screen, such as (although not the same as)

```
TEST CL,3F
JZ 0020
CMP AH,02
JZ 0025
```

You are looking at an example of *assembly code.*

The jumble of numbers was a dump of that part of memory, represented as *hexadecimal bytes.* Believe it or not, this is the most powerful sort of code. By planting the correct bytes, you can cause the computer to execute any instruction of which it is capable.

Since the very early days of computing, programmers have found it much easier to remember *mnemonics* than the raw codes. So CMP stands for "compare," representing hexadecimal 80 (in some cases).

A program called an "assembler" turns the lines of code into the corresponding bytes. It is obviously more productive for a programmer to write code using these mnemonics than to write hexadecimal bytes, but if there is an instruction that the assembler does not "know about," it is impossible to use it. (If you want to be adventurous, type **?** to see a list of instructions including an assembler, or else type **q** to quit and return to normal.)

For more substantial programming, the *compiler* is the method of choice. Code does not now correspond exactly to the *machine code* bytes, but instead represents the mathematical task being attempted. This is where the PC and the simple microprocessor start to part company.

Generations of PC software have become more and more massive, until programs of many megabytes are common. Embedded mechatronic tasks can often be performed by software consisting of only a few thousand bytes, so writing in assembly language is not out of the question. Even though C compilers are available for most microprocessors, the compactness and efficiency of code written in assembler can often be worth the extra effort.

On the other hand, there are very few embeddable microprocessors on which you would want to run the assembler or compiler software to convert your code into bytes. Instead, you will run a *cross-compiler* or *cross-assembler* on a PC to generate bytes that you will download to the microcomputer.

For your programming efforts, you need the home comforts of a keyboard, a mouse, and a screen, plus a disk on which to save the results. The embedded processor is unlikely to have any of these.

There is an alternative to downloading the code bytes to an embedded computer. They can be sent off to the "chip foundry" to be *mask-programmed* into the chip itself. Such chips might cost only a few cents each—but they must be purchased in batches of 10,000!

As I have mentioned, my first embedded computer product, many years ago, used the TMS1000, the chip that powered the very earliest pocket calculators. Its data memory consisted of just sixty-four 4-bit "nibbles"—a nibble is just half a byte. The entire program memory held just over 1000 instructions. Yet this chip had to function as the heart of a cooker clock, counting mains cycles to calculate elapsed time, switching the oven on and off in response to the set cooking time and ready time. It had to show each function on a *vacuum fluorescent display* and to respond to button presses by the user. It had to detect option switches to choose between 50- and 60-Hz mains frequency and between 12-h and 24-h displays. Would you not call that mechatronics?

There was not much fundamental difference between processors of the next generation. You might find the 6502 embedded in a simple controller or else at the heart of an early version of the personal computer such as the BBC Micro or the Apple II. Yet these machines could function as word processors and handle spreadsheets and databases at speeds that seemed hardly slower than those of the modern PC.

Over the years, the 16-bit processor took over the personal computer role, later being outstripped by 32-bit machines and more. Computing power increased by a factor of a 1000, yet software seemed to run no faster.

Most of the power was soaked up by embellishments of the graphics and by an operating system that tried to give the illusion of performing a thousand tasks at the same time. Microsoft realized that for every engineer who needed a machine for computer-aided design, there were a hundred youngsters who wanted a machine on which to play videogames. The Windows operating system has leaned more and more toward providing instant gratification as a music player, a Web-surfing machine, and as a digital television.

It's not all bad.

The move to multimedia has opened up the way to powerful videoprocessing tools. Streams of Webcam images can be captured, dissected with DirectX routines, and analyzed to extract their data. So there are applications where it is preferable to embed an entire PC than to use something simpler. Remember that the "guts" of a PC, minus disk, display, and keyboard, can be purchased as a single board for a very few hundred dollars. A Webcam interfaced to a computer board to process the signal will cost a small fraction of the price of a Pulnix camera alone.

2.5 INTERFACE ELECTRONICS FOR OUTPUT

Almost without exception, the task of getting from a computer's binary output to physical reality will involve responding to a signal that lies between 0 and 5 V. The actual range of output might be only from 0.5 to 3.5 V, and the circuit that it drives must not draw or source a current of more than 1 mA or so.

Something is needed to amplify this signal before we try to use it for moving a motor—or even lighting an LED.

For small signals, the *Darlington driver* has been popular for many years. It amplifies the computer output of a few milliamperes to 0.5 A, enough to drive a small relay or a simple stepper motor. A popular single chip contains eight such drivers, so that 8 output bits of a port such as the printer port could control two stepper motors.

2.5.1 Transistors

The theory of transistors can be ornamented with *hybrid-pi* parameters and matrices, but the essential knowledge for using them is much simpler. Let us start with the *bipolar* or *junction* transistor. This has three connections, a *collector*, a *base*, and an *emitter*. For driving a relay coil, you might connect a transistor as shown in Figure 2.17.

The emitter is connected to ground, the supply is connected to one end of the coil, while the other side of the coil is connected to the collector. As it stands, no current will flow and the coil will be off.

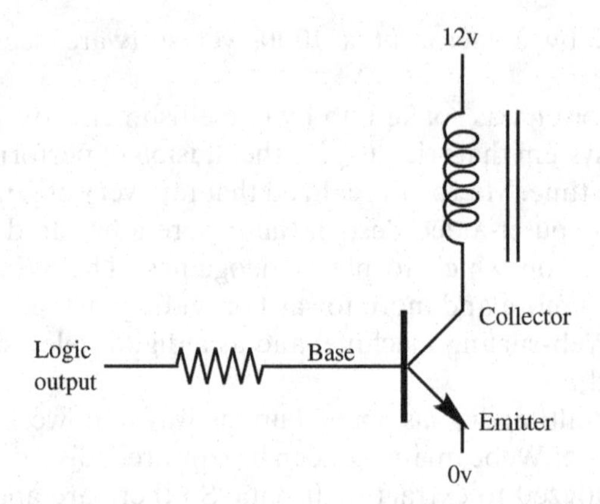

Figure 2.17 *Transistor and coil, with electrodes labeled.*

Now apply some current to the base, to flow through the transistor to the emitter. For every milliampere that is applied, the transistor will allow 100 mA to flow from the collector to the emitter. If it takes 500 mA to turn the relay coil on, we need supply only 5 mA to the base.

Of course, the factor is not exactly 100 for every transistor, but that is a ballpark figure. The actual figure is the *gain* of the transistor, defined by its parameter *beta*.

Not only do we get the current gain of this factor; we get a voltage gain as well. The current entering the base is resisted by a voltage of around 0.7 V. The voltage on the collector for some transistors can be hundreds of volts, but typically you might use 12 or 24 V to drive your relay or stepper motor.

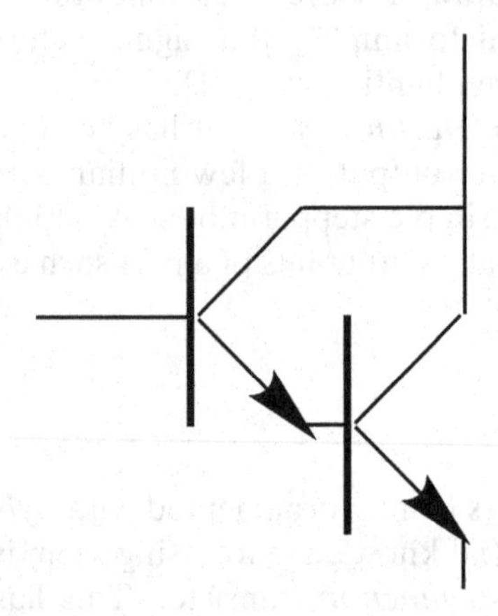

Figure 2.18 *Darlington driver transistor.*

So, if the task can be done with such a simple transistor, why do we bother to use a pair of transistors connected as a Darlington driver (Fig. 2.18)? We must look at the power applied to both the coil and the transistor. Consider a power supply of 12V and a 24Ω coil that will take 0.5A when switched fully on.

When the transistor is off, there is no current and no dissipation. There will be 12V between collector and emitter of the transistor. When the transistor allows 100mA to flow, the voltage across the coil will be $24 \times 0.1 = 2.4$V. That will leave 9.6V across the transistor. With 100mA passing through it, the dissipation in the transistor would be $9.6 \times 0.1 = 0.96$W.

When the transistor allows 250mA to pass, the voltages across coil and transistor will each be 6V. Each will dissipate 1.5W. Overheating of the transistor could certainly be a problem. But when the transistor allows the full 500mA to flow, the voltage on the collector will have dropped to zero and the transistor dissipation will again be zero. To avoid overheating the transistor, we must be sure that it is *saturated*, that we are supplying enough current to the base to pass all the collector current available.

So, in the Darlington configuration, the emitter of the first transistor is connected to the base of the second, while both collectors are connected together. The base current of the second transistor is (1 + beta) times the current applied to the first, so the output current is some 10,000 times the input current—until the collector voltage has fallen too low to supply that much current to the second base. We can be sure that the driver will saturate for a very small input current.

But there are some transistors that can control their output current for no input current at all! These are *field effect* transistors (FETs). There are powerful ones known as *metal oxide semiconductor FETs* (MOSFETs). Once again, the mechatronic engineer need not be concerned with the physics that makes the transistor work, just with the details and pitfalls of using it.

The three connections are now known by entirely different names, *drain*, *source*, and *gate*, but conceptually we can drop the device into the same scenario as the former *junction* or *bipolar* transistor.

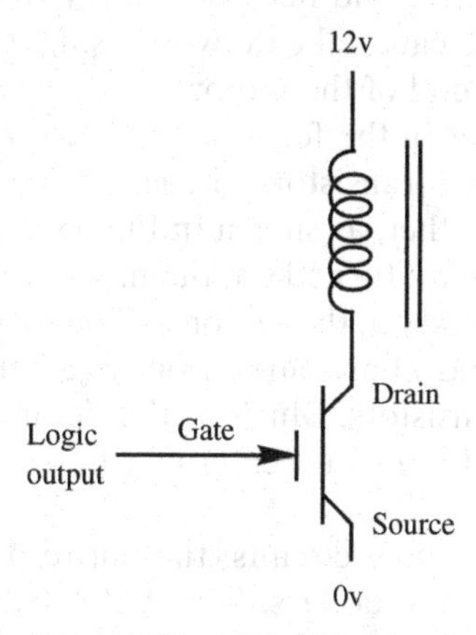

Figure 2.19 *FET driving a coil.*

Connect the source to ground and the relay coil to the drain, and we are ready to control the current with the gate. The essential difference is that when we apply 5 V to the gate, no gate current flows at all! Meanwhile the drain has allowed all the available coil current to flow through the FET to ground, switching the relay coil fully on (see Fig. 2.19).

These devices make interfacing seem all too easy! They have another property that makes us spoiled for choice. Both bipolar transistors and FETs are available in *complementary* forms. N-channel FETs and NPN transistors are used with a positive voltage on the drain or collector and are turned on with a positive gate voltage or a positive base current. P-channel FETs and PNP transistors take a negative voltage on the drain or collector, while they are turned on with a negative gate voltage or base current.

There are several pitfalls to avoid. When a transistor or FET is switched from OFF to ON states, there is a brief burst of dissipation because the output voltage does not drop instantaneously. When it turns off, there is another burst of dissipation that can last somewhat longer. In general, this is not serious. If the controlling computer switches the drive on and off repeatedly at high speed, however, dissipation can become a problem.

A second problem can result from the inductance of the load, whether it is a coil or motor winding. As the current is turned off, there is a spike of voltage in a direction that tries to keep the current flowing. The use of a diode across the coil will avoid this voltage, but the current in the coil will fall more gradually.

So far we have considered simple on/off switching, but in the case of a DC motor we usually want to reverse the direction of the current in the motor. For that, a convenient solution is the H-bridge.

2.5.2 The H-Bridge

If we connect one end of the motor to ground, we would have to provide current from both a positive and negative supply to ensure two-way control of the motor. Instead, we can make do with a single supply if we can switch the connection of either end of the motor.

This results in a circuit in the form of an H, with the motor representing the crossbar and with two transistors in each of the "uprights": A and B in one and C and D in the other, as shown in Figure 2.20.

If transistors A and D are turned on, the motor will be driven one way. If, instead, B and C are turned on, the motor will be driven in the reverse direction. If A and B or C and D are turned on together, however, it will mean instant death for the transistors. Much of the detail of any H-bridge circuit (see example configuration in Fig. 2.21) will be there to prevent this from happening.

As with so many electronic circuits, the entire device can be purchased "on a chip" for a very few dollars. The L298 from ST Microelectronics (http://www.st.com, see also http://www.learn-c.com/l298.pdf) contains two

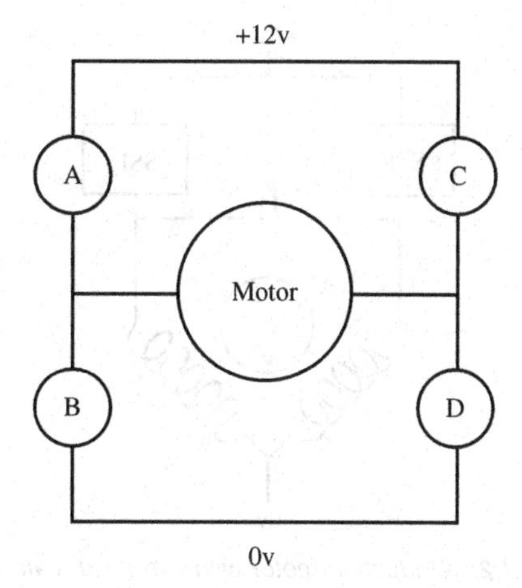

Figure 2.20 H-bridge.

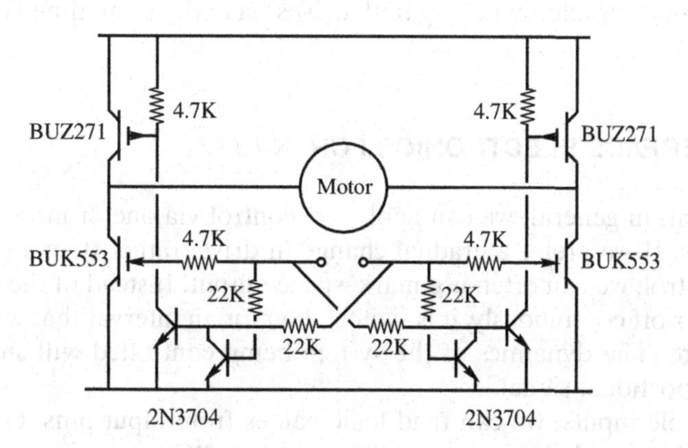

Figure 2.21 H-bridge circuit.

H-bridge circuits and will drive small motors with ease. For larger loads, however, you might need to design and build a more specialized circuit.

2.5.3 The Solid-State Relay

Another useful on-a-chip device is the solid-state relay. It contains a light-emitting diode and a photosensitive triac, with the result that a single output bit of the computer can switch an AC device on or off. The LED provides complete electrical isolation between the computer and the AC circuit.

Figure 2.22 shows an example in which a two-phase induction motor can be driven forward or in reverse. It was used with great success to automate the operation of the valves of a pilot sewage plant! Time has to be allowed for the motor to stop before reversing it, but the control was very leisurely.

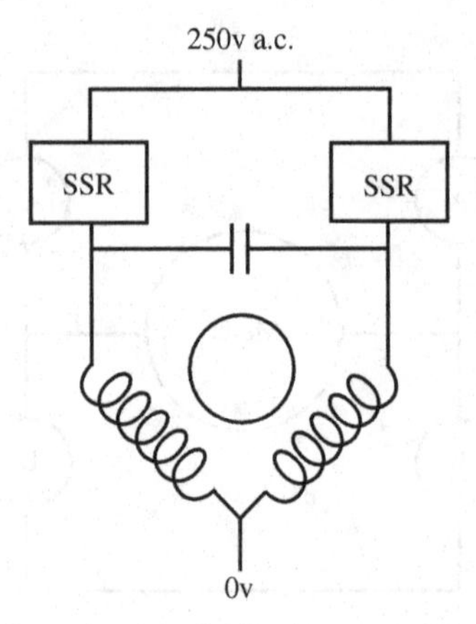

Figure 2.22 *Induction motor with two solid-state relays.*

Although undesirable, switching both drives on at the same time need not be disastrous.

2.6 INTERFACE ELECTRONICS FOR INPUT

We see that, in general, we can apply our control via one or more computer output bits. If we prefer a gradual change in drive, rather than a *bang-bang* on/off control, we can often use mark–space output. Instead of the output bit being on or off continuously, it is switched on for an interval that we can vary by software. The dynamics of the system being controlled will smooth this into a proportional signal.

For simple inputs, we can read logic values from input pins, both on the simplest of embedded microcontrollers and on a PC used as a controller.

Reading analog sensor values into the computer is a different matter. Now we can have a continuously varying voltage, such as a tachometer or potentiometer output, that we want the computer to be able to read to considerable precision. We need some means to convert this analogue signal to digital form.

2.6.1 Simple Inputs

Perhaps our inputs are merely logic signals. We can then arrange for them to pull one of our logic inputs to ground, maybe using a transistor to avoid applying the signal directly to a computer pin. If we are particularly concerned about the computer's safety, we can use an *opto isolator* (Fig. 2.23). The package, usually containing four channels or more, has a LED activated by the input signal and a phototransistor that conducts when the LED is lit. In this way, there is no electrical connection between the computer and the sensor circuit.

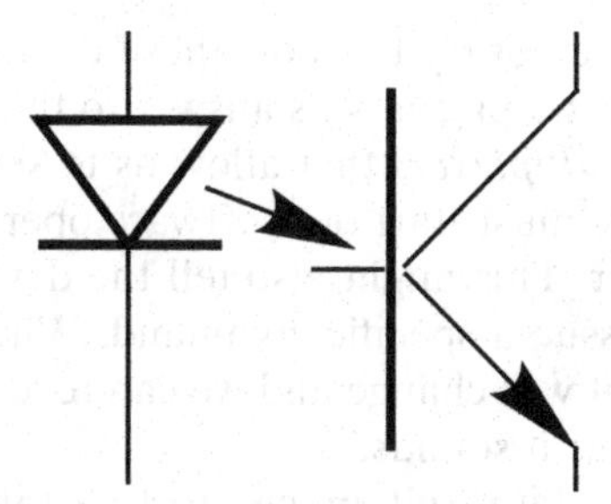

Figure 2.23 Opto isolator.

The printer port of the PC, in addition to its eight output data lines, has five logic inputs. Indeed, a single output command can also convert those eight output lines to inputs.

2.6.2 The Analog-to-Digital Converter

For reading an analog sensor, we need something more.

Analog-to-digital converters (ADCs) come in a variety of forms, from the very simple circuit that reads the movement of a game joystick to sophisticated systems that can "grab" waveforms at video speeds and above and pack the data into a block of memory.

The vast choice is one of the problems that beset the mechatronics system designer.

At the simplest level, a ramping voltage can be compared against the signal to be measured. The resulting number is obtained from a counter that measures the time taken for the ramp to reach the signal voltage. It is simple, but it is slow if any sort of precision is desired. It can be constructed from a few dollars' worth of components to connect to the printer port, if no other ADC can be found. A more detailed design is given in Section 5.3.5, together with simple software needed to drive it.

Many converters are built around a digital-to-analog converter (DAC), the counterpart of the ADC, that converts a number to an analog output. With this, it is possible to use a binary search method, termed *successive approximation.*

Suppose that the input can be in a range of 0–8 V and that the actual value is 6.5 V. First, the DAC output is set to 4 V, the halfway point, and a comparator circuit gives the answer to the question "Bigger or smaller?" As 6.5 V is bigger, so the next DAC value is 6 V, three-quarters or 75% of the maximum. The answer is still "bigger," so the next ADC value is 7 V, dividing the 6–8 V range in half. Now the answer is "smaller."

The sequence "bigger–bigger–smaller" becomes a binary number 110 that defines the value we are looking for. By successively dividing the range to search in half, a precision of 1 part in 4000 can be gained from just 12 tests.

In general, it is not necessary to know these details. A circuit deals with all the logic operations and presents its answer to the computer.

Many ADCs have *multiplexers* that allow us to select between a number of input channels. So we must start our software operation by outputting the desired channel number. This might also tell the device to start the conversion, or else we must issue a specific command. When the answer is ready, the status of an input bit will change and we can read the data. However this might not be as simple as it sounds.

If the ADC gives a 12-bit result, precise to 1 part in 4000, and if our input is an 8-bit byte, we have to find a way to read the answer in two parts and to combine these together. Rather than go into details here, see the code examples in Chapter 3.

For very high-speed data acquisition, there is the *flash converter*, which relies on brute strength and a large number of comparators to give an answer in a small fraction of a microsecond, so that video signals can be encoded.

Commercial interface cards are becoming more and more complicated—and more expensive, too. They could be very simple, indeed, containing little more than a single-chip ADC that can be purchased with 16 input channels for under $5. Instead, they often contain first-in first-out (FIFO) buffering for 1000 samples or offer conversion rates of many megasamples per second.

The ADCs that could be purchased in the late 1990s were simple and ideal for student interfacing, requiring no more than a dozen lines of code to drive them. There has been a more recent trend for cards to hide behind a megabyte of driver software and to expect the user to depend on some software environment such as Simulink. Nevertheless, a close look at the manual will reveal that they have the same essential structure.

It is unfortunate that the ADCs of the "sound card" cannot easily be exploited for online control. Found in just about every PC, they are designed to digitize the two channels of stereo audio at 44kHz. However, the audio signals are *AC-coupled*, meaning that steady voltages are lost.

Nevertheless, for laboratory experiments there is an alternative. The simple chip, the MCP3204 from Microchip Technology Inc., provides four channels of 12-bit ADC conversion for a few dollars. It communicates serially, in a way that can be connected directly to the parallel port of a PC with no other components whatsoever. It is described fully in Chapter 5, Section 5.3.4.

For another alternative, it is no great task to attach an embeddable microprocessor such as a PIC chip or the Motorola HC12 to the serial port of a PC. The tasks of ADC and control outputs can then be delegated to the microcontroller, at the cost of some slight delay in the data communication. Indeed, the entire control task can be performed by the microcontroller, but for software development or laboratory data logging, the home comforts of the PC remain very attractive.

Mark Phythian has designed just such a PIC-based ADC 4-bit output as well. Full details of the circuit and software are given in Section 11.3.1.

2.6.3 Other Inputs

In the case of a PC, we can, of course, use the computer's more general inputs for sensor data. The older serial computer mouse connects to a COM port that receives signals encoded in ASCII format. Inside the mouse, a chip does all the necessary decoding and analysis of the sensors that detect its movement and sends trains of 5 bytes at 1200 Hz to the serial port of the computer. Other versions of the mouse can be connected to a PS2 or USB port, but the principle is the same. The mouse chip does the hard work and sends processed data to the computer.

By reading and interpreting the mouse codes directly, we can bypass the acceleration that would prevent us from using them for absolute displacement measurement. For a few dollars, we have two channels of position sensor with resolutions of 0.1 mm.

Once we have delegated tasks to other devices, the choices become even more bewildering. Communication systems such as CANBUS are available at chip level and allow us to connect a whole network of devices together. We can use these devices as *autonomous agents*. We should not rule out Ethernet and other general networking systems, where protocols such as TCP/IP can allow devices to call each other up with all the ease of browsing internet pages.

2.6.4 Signal Conditioning and the Operational Amplifier

We may have a signal that is not suitable for connection to an ADC, possibly because the voltage is too small. The *operational amplifier* (Fig. 2.24) comes to the rescue.

The design of a linear transistor amplifier can be a tricky business, with the need to provide the correct bias currents and to match the gains of transistors to minimize the drift effects of changing temperatures. Fortunately, the makers of integrated circuits addressed this problem many decades ago. The operational amplifier sells for a few cents and can be configured to perform a wide range of tasks.

Apart from two power supply connections, in essence it has just three terminals: the output, the inverting input, and the noninverting input. It can be described by just one equation

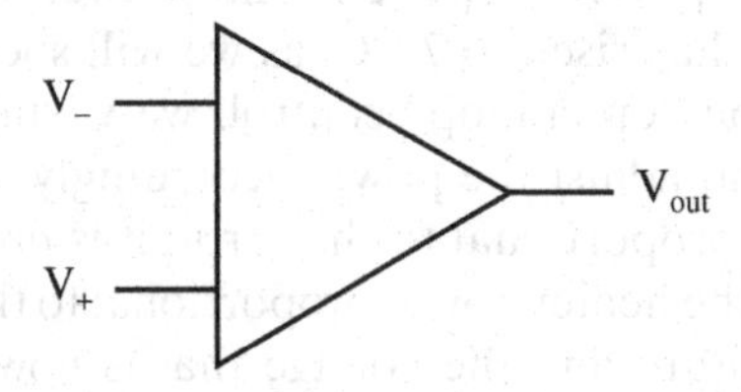

Figure 2.24 *Schematic representation of an operational amplifier.*

$$V_{out} = A(V_+ - V_-)$$

where A is a very large number, typically 100,000. Some earlier designs had additional connections for frequency compensation, but such considerations are unnecessary now except for unusual applications.

It is amazing that such a straightforward device can come in so many varieties. Some have more powerful output, some have exceptionally small drift, while others have an extended frequency response or rapid "slew rate."

A feature to look for is the ability to work between supply rails of 0 and 5V (many are at their best with +/– 15V) with an output voltage that can swing rail to rail.

Circuit design techniques are given in Chapter 5.

2.7 PRAGMATIC CONTROL

Control theory can be a mathematician's delight, blending *calculus of variations* and the *two-endpoint problem* with all things fuzzy, neural, genetic, chaotic, and catastrophic, with a generous helping of complex variables and matrices thrown in. In practice, the essence of designing a controller is an understanding of the dynamic behavior of the system to be controlled.

2.7.1 The PID Controller

The three terms in the three-term or "PID" controller are *proportional*, *integral*, and *derivative*. Over the decades its principles have been applied, starting with mechanical contraptions with ball-and-plate integrators or pneumatic devices worked by air power, all the way to the latest of the *Programmable Logic Controllers* sold as a universal cure for control problems.

We must start with the fundamental concept of *feedback*. First consider a system without it. The temperature of a directly heated bathroom shower could be controlled by a simple power controller wired to its electric heater. As the power is increased, the flowing water gets hotter. The knob of the power controller could be calibrated in degrees and the job would be done. At the mark on the knob labeled 45°C, let us suppose that 50% of full power is applied to give the desired temperature.

But what happens when a tap is turned on in the kitchen? The flow decreases by half while the same power is applied. The shower user lets out a scalded yelp. The temperature has risen to 70°C, as we will soon see.

Instead of relying on "open loop" control, we would like to measure the output temperature and adjust the power accordingly. This is feedback, and if we make the power proportional to the error, it is *proportional feedback*.

So now let us make the heater power proportional to the difference between the measured temperature and the voltage that is now given by the setting knob:

- Suppose that full power gives a rise in temperature of 50°C, so that each 1% increase in power will give us a rise of 0.5°C.
- Suppose also that the cold-water supply is at 20°C.
- Suppose that the target setting is at 45°C.
- Suppose that the *gain*, the factor by which we multiply the error to get the power setting, is G percent heating per degree of error.

Then, if the output temperature is T degrees, the percentage of power applied will be

$$(45 - T)G$$

resulting in an output temperature

$$20 + 0.5(45 - T)G$$

Now we can equate this to the temperature T, to get

$$T = 20 + 0.5(45 - T)G$$

which we can solve to get

$$T = (20 + 0.5G45)/(1 + 0.5G)$$

If $G = 2$, this gives an output temperature of (20 + 45)/2 or 32.5°C, a long way below the target. If $G = 20$, we get (20 + 450)/11 or 42.8°C, much closer.

But to see the real advantage of feedback, consider the effect of halving the flow. Now at half-flow, full power gives 100°C temperature rise. So if we relied on open-loop control applying half-power, we would get an output temperature of 70°C.

With closed-loop control and $G = 2$ we would get (20 + 90)/3 = 36.7°C. Turning on the kitchen tap has caused an increase of 4.2°C, which is much safer.

With $G = 20$, we get (20 + 900)/21 = 43.8°C, a jump of only one degree.

So, feedback can greatly reduce the effect of a disturbance, and as the feedback gain is increased, the error away from the target gets smaller.

Why not make the gain infinite? The *infinite-gain controller* can take the form of a simple thermostat switch that cuts the power as the temperature rises above 45°C. As the heater starts to cool, the temperature drops below 44°C, say, at which the system switches the power on again.

We could complain that the temperature is oscillating as the switch opens and closes in a *limit cycle*, but the one-degree variation is probably quite acceptable. Indeed, we could say with some justification that at least 90% of all control systems are unstable and oscillate, since that is probably the proportion of thermostats and similar controllers in everyday use.

Infinite gain is an option in this system, where the temperature can be measured very close to the heater, but there are many more systems where it is not. These may require a "smooth" output. Perhaps the necessary sensor signals with "immediate response" are not available. For numerous reasons, we may have to use more ingenuity in the controller.

The *proportional* feedback gain G accounts for the P in PID, but what of the other letters? If we were forced to measure the shower temperature further downstream, the oscillation might become rather serious. Oscillation in the shower temperature can often occur with human control. If the temperature is low, we increase the power. But the water, now too hot, flows through the hose and does not hit our skin sensor until a second or two later. When it does, we cut the power and suffer the further second of overly hot water, only to be hit by a chilling blast.

The solution is to reduce the size of our adjustment and inch the setting slowly up, waiting for the effect before making each new adjustment.

Consider an automatic temperature control system that oscillates unless we reduce the gain considerably, say, to 2 or below. Now the temperature error is larger than we can accept. How do we correct it? We can "integrate" the error, to get a term I that increases with time and with the error—this is *integral feedback.*

If we add this to the power drive, we will have an output temperature

$$20 + 0.5((45 - T)G + I)$$

or after solving with $G = 2$

$$\begin{aligned} T &= (20 + 0.5G45 + 0.5I)/(1 + 0.5G) \\ &= (20 + 45 + 0.5*I)/2 \end{aligned}$$

The temperature starts at first to settle at 32.5°C, but all the time it is below the target of 45°C, the value of I will increase steadily—although not so fast that it will cause oscillation. After a time (theoretically infinite), I will reach the value 50, which will cause the output temperature to be at 45°C—perfect!

But there is a catch. Suppose that the kitchen tap is turned on and the flowrate halves when the integrator has reached this value of 50. Now, with the flow halved, the 0.5 becomes 1.0, so the new equation for the temperature is

$$\begin{aligned} T &= (20 + 90 + I)/(1 + 2) \\ &= (20 + 90 + 50)/3 \text{ since } I \text{ has not had time to change} \\ &= 53.3 \text{ degrees} \end{aligned}$$

which is probably still enough rise to cause a yelp of pain.

Integral control is a valuable addition for tuning a process plant to give a steady exact output, but is less useful when large disturbances can occur suddenly. That leaves us with the task of explaining *derivative feedback*.

Suppose that we are trying to control a motor by means of an input u that sets its acceleration. The differential equation that describes the motor is

$$\frac{d^2x}{dt^2} = -u$$

If we apply a feedback signal proportional to x, we have

$$\frac{d^2x}{dt^2} = -kx$$

But as we will see, this is the same equation that we get for a pendulum. For any positive value of k, the result will be an oscillation. (For any negative value, it will cause x to run away toward infinity!)

The D term in the controller estimates a derivative of x (assuming that we cannot actually measure one), and we will later see that it can make the system stable.

2.7.2 Understanding the Dynamics of a System

The theory taught in many courses is centered on the manipulation of transfer functions. But a real problem does not start with a transfer function. Instead, you are confronted by a physical system for which you have to derive the differential equations yourself before any kind of mathematical analysis can even start. What is more, the equations will certainly have some serious nonlinearities, such as the "full drive" limit of the acceleration of a motor or the "bottoming" of a spring shock absorber.

The most valuable contribution of "modern control theory" (well, it's still only half a century old) is the concept of the system's *state*. This is a set of properties such as position and velocity that define exactly what it is doing at any instant.

Let us consider a classical second-order system, a swinging pendulum. The acceleration of the bob is proportional to its displacement and is directed toward the center. In equation form, this becomes

$$\frac{d^2x}{dt^2} = -n^2x$$

If we differentiate $\cos(nt)$ twice, we get $-n^2\cos(nt)$. Similarly, if we differentiate $\sin(nt)$ twice, we get $-n^2\sin(nt)$ The general solution is a mixture of sines and cosines of nt

$$A\cos(nt) + B\sin(nt)$$

with coefficients A and B that are determined by the *initial conditions*. These initial conditions will define the motion from that time on.

But what if we had measured the values of the initial conditions a moment later? The solution must still be the same. We can define any time throughout the swinging as our *start time* and use those initial conditions to define the future behavior of the pendulum.

So, those initial conditions are no longer quite so "initial," but carry forward all that we need to know about the past movement, in order to predict the future. Those initial conditions have become "state variables."

Each such variable has a rate of change that depends only on itself and other state variables, or on the input of the system if it has one. Our two variables in this case are the position and velocity of the bob, x and v.

The first rate-of-change equation is rather obvious:

$$\frac{dx}{dt} = v$$

The second equates the rate of change of v to the acceleration:

$$\frac{dv}{dt} = -n^2 x$$

Instead of a second-order equation, we have two first-order equations.

For a simple position controller, there will similarly be two state variables, position and velocity. The first equation appears almost trivial—it just states that rate of change of position is equal to the velocity.

The second equation concerns the rate of change of the velocity, and will express it in terms of the input drive and the velocity itself if there is any kind of friction. Now we can concern ourselves with the way in which feedback will change the dynamics of the system.

Having worked out the equations, it is a matter of simplicity to write a few lines of computer code to simulate the system, nonlinearities and all. It is easy to add a few more lines to represent a control strategy and deduce the response of the closed-loop system.

If these methods look a bit cut and try, they can be given a mathematical gloss by using matrices to represent the equations when they are linear. The matrices can be manipulated to represent the closed-loop system, too. In a later section we will see that a routine mathematical operation gives an equation, the characteristic equation, from which the stability of the system can be assessed.

But that is getting ahead of the practicalities of designing a controller. Let us consider, for example, the control of the end effector of a robot.

2.7.3 Practical Design

The first task of the designer will have been to choose a motor and gearbox that will give an appropriate top speed and torque. Let us assume that sensors are in place to measure both the position and the velocity.

It can be shown that a simple second-order system such as this will be stable for any values of negative feedback of the two variables. But mere stability is not enough. We want a response that does not overshoot and a final position that is not deflected excessively if there are disturbing forces.

Linear theory assumes that if the error is doubled, the drive to the motor will also be doubled. But this is a real motor, with a real drive amplifier that has a limit on the output it can give. To be sure that the error is kept within bounds, we must ensure that this full drive is applied if the maximum acceptable error is exceeded. This error is termed the *proportional band*, the range of steady errors for which the drive will be proportional to the error. Outside this band, the drive will saturate.

How much velocity should we feed back? The simple answer is: "Enough to avoid an overshoot." From the largest likely initial error, the motor must accelerate toward the target, but start to decelerate soon enough that the load can come to rest without overshoot. As the speed builds up, the contribution of the velocity feedback must cancel out the position term by the halfway point.

Pragmatic design such as this is greatly at odds with the linear approach that is conventionally taught. But when the system is dominated by a nonlinearity such as drive saturation, the linear approach is no longer appropriate. That is not to say that we cannot apply analytic tools to the task, as you will see later in the book.

2.8 ROBOTICS AND KINEMATICS

Machine tools were traditionally built around slideways and pivots, so that in a lathe the work was spun about an axis while a cutter moved in straight lines to carve it into cylindrical sections. Generally, just one of the controlled axes would be adjusted at a time. The involvement of a computer has meant that robot manipulators could be freed from the need to move in straight lines and circles.

Now we are faced with a robot with six rotary joints, linking arm segments in series in a way that can in theory reach any point in its workspace with a toolpiece held at any angle.

The art of calculating the *end effector* position and orientation from the six angles of the axes is termed *kinematics*. Some elegant matrix methods that we will meet later have been developed to ease the calculations. But that is only half the problem—and the easy half at that.

If we know where we want to put the tool, we need a way of calculating the axis angles that will achieve it. This is a much less straightforward calculation called *inverse kinematics*. We will also be interested in the relationship between the velocities of the axis angles and the velocity of the toolpiece. This is where *dynamics* comes in.

For all except the simplest of mechanisms, these problems are likely to get mathematically intense. This is a good place at which to round off the bare essentials and dig deeper into the toolbox.

3

Gaining Experience

These initial remarks are directed to the lecturer or examiner of a course on mechatronics. The experiment instructions that follow can be used as course notes for the students.

Practical laboratory experience is an essential ingredient for linking together the diverse aspects of mechatronics. But it is necessary to choose among a wealth of alternatives when selecting and designing experiments.

Numerous experiments are available on the market, but they are usually very costly. Much worse, many of them require very little creativity or understanding on the part of the student, consisting instead of a ritual of knob turning, measurement, and graph plotting.

In the experiments that follow, the students are required to create software from scratch, not just by dragging icons in a Graphic User Interface. They add hardware by connecting circuits that they could easily replicate from component level—indeed, an able technician should have no difficulty in fabricating these experiments from catalog components.

There is, however, the problem of the choice of computing platform.

When mechatronics ability comes to be applied by the graduated student, will it be to use a PC for control and coordination tasks, or will the objective be a mechatronic product with an embedded microcomputer? Should the software lean toward the latest version of the .NET environment, or should it be based on downloading the simplest code to a single chip?

An embeddable chip such as the HC12 certainly has all the input–output capability and the computing power to control any of the experiments here.

It has numerous ADCs and output pins, with an abundance of program memory. But its "user interface" is limited, to say the least. It will need to be linked via its RS232 serial capability to a PC (or its equivalent) on which the student can write, edit, save and then compile or assemble the software. The link will also be needed to upload data for display or plotting.

Instead, a simple PC can in principle perform all the tasks of development interface, performance monitor, and real-time controller, with no additional circuitry beyond an output driver to power any motors and a simple analog-to-digital converter for reading sensor signals. The fly in the ointment is that simple PCs and ADCs are becoming increasingly hard to find!

If you do not have a suitable ADC card, you will find the software in Chapter 5 for using an MCP3204 four-channel 12-bit ADC chip, costing only a few dollars. Apart from a connector for the printer port and a length of ribbon cable, no other components are needed for the interface if the signal voltages are of low impedance. The chip can even be powered from one of the output pins. One more chip such as the TL074 can provide four buffer amplifiers for higher-impedance signals.

Each elaboration of the Windows operating system seems designed to put more distance between the user and the actual operation of the hardware. Vendors have dropped simple ADC cards from their lists in favor of FIFO-buffered elaborations that require a monstrous software driver library to use. However there are "workarounds" to enable you to use even these, although life is much easier if you have a board of traditional design. Also, many problems can be avoided if you have retained a copy of your old Windows 98.

An alternative is to use the HC12 or a similar microprocessor as a "slave" interface. With a simple protocol, it can be asked to reply with the values on any of the ADC inputs, to latch output values and apply them to output bits or mark–space registers. Using the serial interface of the PC, rather than a proprietary plugged-in card, it should be proof for some years from the efforts of the operating system writers. The effective conversion time will be greatly increased by the serial communication, but when set to a high baud rate, this should still be acceptable.

This might at first threaten to present the same dangers as the Labview approach, in which the interfacing is regarded as a piece of magic into which the student is not supposed to delve. But the student can certainly unravel the simple code of the HC12 to see how it ticks. What is more, a second level of the experiments can see the students writing HC12 code to achieve the control objectives without the intervention of the PC, once the code has been downloaded.

A further resource, to be found on the accompanying Website at http://www.essmech.com/3/vb.htm, is an example of the use of Visual Basic rather than QBasic for constructing code for these experiments. The conversion is a very simple one.

The first experiment, however, requires no input at all. It uses the printer port to drive two stepper motors that will form the essence of a "mobile robot." There is little theory involved, but it offers the great satisfaction of seeing a computer moving a "robot" about with homegrown software. But first it is necessary to settle on the software environment to use. The following discussion is aimed at both instructors and students.

3.1 COMING TO GRIPS WITH QBasic

The choice of computer language is always a thorny issue. Partisan support can be as ardent as that of any football supporter, and any choice is going to upset somebody. It could be tempting to select the "newest and the best," perhaps a version of C with the greatest number of pluses after it, but for getting started the greatest simplicity will give the greatest advantage.

An early mainstream language was FORTRAN, but it was hidebound with conventions designed to fit in with the use of punchcards! Some time later it was followed by Algol, a language that treated line endings with disdain and abounded in semicolons.

When "personal" microcomputers began to be sold, the advantage of simplicity had market value. Two varieties of Basic, BasicA, and GWBasic had simple syntax and were well within the capabilities of a generation of schoolchildren. However, they had their roots in "line at a time" program entry and editing and depended heavily on line numbers, even more so than FORTRAN.

Meanwhile Algol had evolved into Pascal, taking its semicolons with it. Algol and Pascal both use labels, rather than numbers, to tag special points in their code.

Soon on-screen editing was the only way to go, and line numbers could be dropped. Quick Basic borrowed some of the best features of Pascal and combined them with the simplicity of Basic. Version 4.5 included a compiler that could efficiently reduce your code to an .EXE file and this was soon followed by Visual Basic for DOS.

Quick Basic caught the attention of Microsoft and a stripped-down version, QBasic, was included in DOS operating system disks and in installation disks for all versions of Windows before XP. Meanwhile, the same syntax was used for Visual Basic, both as part of Visual Studio and as a scripting language in most, if not all, Microsoft Office applications.

It is probable that any serious real-time programming will be performed in some version of C. It is second cousin to assembly language, the lowest level at which it is convenient to program a microcomputer, and as such it has access to processes at a fundamental level. However, C has quirks of cryptic syntax that make reading the software of even the most careful programmer an arduous exercise in concentration.

QBasic code can be read like a novel, as I hope that you will soon agree. Just as in C, there are ways to perform fundamental operations such as writing and reading bytes directly to or from peripheral interfaces.

3.1.1 A Simple Start

Launch QBasic, and the introductory page will appear. You can take a quick tour of the Help files or press ⟨escape⟩ to go straight in. Type in the program line

```
play "cdeccdecefgn0efg"
```

As you press ⟨return⟩ at the end, you will see the word PLAY change into uppercase. It means that the syntax has been checked and the keyword recognized.

Now run the program—yes, it really is a program. There are two ways. You can click on RUN in the menu bar and then START in the submenu that drops down, or else you can press ⟨shift–F5⟩.

What did it do? You should have heard a simple tune.

If you want to know more about the PLAY routine, put the cursor on the word and press ⟨F1⟩—there's everything you could ever ask for. Press ⟨escape⟩ to clear the Help page.

So, what has this got to do with control?

Here we have a very simple way to measure out time, because the music plays the notes at a speed that does not depend on the speed of the processor. We can easily control the speed at which it plays—put another line at the start of the program, to get

```
PLAY "L8"
PLAY "cdeccdecefgn0efg"
```

Run it again. What is the difference? Now change `L8` to `L16`, run again. Try `L64`. Then try `"L64T255"`.

So, a PLAY statement can measure out an interval of time that is independent of the speed of the computer. In order to perform real-time digital filtering, we have to have good control of timing.

3.1.2 Using Graphics

Clear your program and start again by selecting NEW. (To find it, click on FILE.) Do not save.

The line

```
SCREEN 12
```

will set a graphics mode. A second line

```
WINDOW (0, -1.1) - (1000, 1.1)
```

sets the screen coordinates to a range of 0–1000 across and –1.1 to +1.1 from bottom to top. Then

```
LINE (0, 0) - (1000, 0), 9
```

It will draw a blue line across the screen. The `9` specifies "bright blue." Put the cursor on the word LINE and press ⟨F1⟩ to see all the details.

Put in these three lines and run the program. Well, it's a start! Now let's draw something. Enter the following lines:

```
FOR i = 1 TO 1000
  PSET (i, SIN(I / 100)
NEXT
```

The PSET (point set) routine puts a dot at the coordinates in the brackets.

Run the program, and a sine wave will flash onto the screen. So how do we slow it down? Put the line

```
PLAY "L64T255"
```

at the start of the program and

```
  PLAY "n0"
```

just before the `NEXT` line. (`n0` is the code for silence!)

Now you see the sine wave crawl across the screen, taking 12 seconds to run.

3.1.3 A Real-Time Model

Let us now try to model a lowpass system—an example of such a system would be a resistor–capacitor *lag*. It is really not difficult, although you will find much more information on the theory later in the book.

First we need an input signal. Let us make a square wave. A neat way to do it is with a logic operation.

In numbers ranging from 128 to 255, from 384 to 511, and two more ranges below 1000, the "128-bit" of the binary value is set. So the logic function `(i AND 128)` will cycle from 0 to 128 four times.

```
IF (i AND 128)>0 THEN
  u = 1
ELSE
  u = -1
END IF
```

Put these lines just before the `PSET` line and change the `PSET` to

```
PSET(i, u)
```

This time you should see four cycles of square-wave—lines of dots at the top and the bottom of the screen.

So, how do we make the lowpass filter? Try adding the lines

```
x = x + (u - x) / 20
PSET (i, x), 14
```

after the `END IF` line. There in yellow is the waveform that you would expect from an *RC* circuit driven by a square wave.

So, how does it work? The differential equation corresponding to a lowpass filter with time constant T and input u is

$$T\,dx/dt = u - x$$

So to a first (and pretty good) approximation, the change in x over a small time dt is

$$(u - x)\,dt/T$$

or in code terms

```
x = x + (u - x) * dt / T
```

Instead of dt/T, we have used the numeric value 1/20 in the code above—in other words, T has the value $20\,dt$. Now in our real-time plot, we have dt = 12 ms, so the time constant T is 240 ms or around a quarter of a second.

Later we will see how this sort of filter can be useful.

You might like to save this as SIM1.BAS before you choose NEW.

3.1.4 SUBs and FUNCTIONs

Clear the decks again with NEW (first click on FILE).

By defining a function, a single word can be put in your program to represent a whole operation such as reading an input device. Type the line

```
function twice(n)
```

On pressing ⟨return⟩ on your keyboard, you will see this change to

```
FUNCTION twice(n)
END FUNCTION
```

and now you can type in your function between these lines. Put in the line

```
twice = 2 * n
```

Now we are on a special page just devoted to this function. To get up to the main program, press ⟨F2⟩—where you will see the top line with UNTITLED and a second line with TWICE. Click on UNTITLED. You will see a blank page. Now enter the line

```
PRINT twice(7)
```

and run the program. You should not be too surprised to see the number 14 appear.

Functions can be called recursively. Try changing the line to

```
PRINT twice(twice(7))
```

and run the program.

SUBs are similar, except that they do not return a value. Clear the decks again and type

```
SUB treble(n)
n = 3 * n
```

(The computer will have added an END SUB.)

Press ⟨F2⟩ to go to the main program page (UNTITLED). Now enter the lines

```
CLS 'Clear the screen
x = 5
treble x
PRINT "The answer is "; x
```

There are a few points to notice. In the SUB, n is a "dummy variable." The SUB does its task on whatever variables are passed to it, in this case the variable x. The value of x is changed inside the routine. (For C buffs, the default is that x is passed as a pointer)

The single quote after CLS means that the rest of the line is a "remark" and is not treated as code.

3.2 THE SIMPLEST MOBILE ROBOT

The "turtle" was popular in the mid-1980s for teaching children the rudiments of programming. It accepted combinations of commands telling it to turn or advance, then it trundled across the floor.

In essence, it consisted of two stepper motors, one driving the left wheel and the other, the right. It steered like a wheelchair, by turning the driving wheels by differing amounts, while skids in front and behind stopped it toppling.

3.2.1 Driving a single Stepper

A starting point for this project is to drive just a single stepper motor. Raid the junk heap for a discarded $5\frac{1}{4}$-in. floppy disk drive. In addition to the electronics and a very interesting motor that rotates the disk, it has a square, chunky stepper motor that drives the head in and out.

This stepper has the advantage that it will operate on a small current, and if you are prepared to take the risk (use a "thirdhand" computer), you can drive it directly from the printer port. It has the disadvantage that when operated in this way it has very low torque, and serves merely to demonstrate how a stepper steps. Its two windings have no center tap, so to use it in the "turtle," you would have to use an H-bridge driver rather than the simpler Darlington driver. But more of that later.

Use a test meter to determine how the four leads are connected to the two windings—pair those leads between which you find some conduction as NS and EW.

To use the printer port, you will need a printer cable. Crimp a pair of 25-way connectors onto a ribbon cable, so that you have plugs of opposite "genders" at the ends, wired pin to corresponding pin. This will act as a printer port extender cable. When connected, the "pins" at the free end will take the form of hollow sockets.

For this elementary test, you can push wires into the connector sockets to attach the motor—connect N, S, E, and W to sockets 2, 3, 4, and 5, respectively.

You will need to know the address of the printer port—it is probably at `&H378` as listed in the software below, but might instead be at `&H278`. The `&H` means that the rest of the number is in the hexadecimal scale of 16. So `&H10` is the decimal value 16, `&H100` = 256, and `&HFF` = 255.

You can use Windows to find the address of the port. Follow the trail START, SETTINGS, CONTROL PANEL, SYSTEM, DEVICE MANAGER, PORTS, LPT1, PROPERTIES, and RESOURCES—and there at long last you will see the address range of the port. The first address is the one you want. Substitute your correct address in all the statements below.

Launch QBasic. Then twiddle the motor shaft with your finger and thumb. It should turn freely.

Enter the following line of code, and then run it:

```
OUT &H378, 1
```

Feel the motor again—it should feel lumpy when you turn it.

You have just applied 5 V (or a little less) across the NS winding. Try

```
OUT &H378, 5
```

and run it. The lumpiness should be somewhat greater—you have applied 5 V across both the NS and EW windings.

For neatness, run

```
OUT &H378, 0
```

to ease the load on the printer port.

Stick and fold a label over the motor shaft, to form a pointer so that you can see any movement more clearly.

Now is a good time to start to give some structure to the code. Let us first define the port address as a constant, in the form

```
CONST port = &H378
```

Now we can arrange the main part of the program as a loop, as follows:

```
FOR i = 1 TO 50
   stepto 5
   stepto 6
   stepto 10
   stepto 9
NEXT
```

So, what does stepto do? It does nothing until we write it!

```
SUB stepto(n)
   OUT port, n
   PLAY n0
END SUB
```

There's another use for the PLAY routine for timing.

Run the program, and you should see the pointer stepping sedately round, making one complete revolution if the motor has 200 steps per revolution. Feel the torque that is required to stop it.

Now speed things up by adding the line

```
PLAY "L64T255"
```

at the top of the program.

Rotation should now be much more brisk. But what has happened to the torque, when you grasp the shaft?

The significance of the numbers 5, 6, 10, and 9 is that they represent the binary numbers

```
0101
0110
1010
1001
```

The 4-bit code can be regarded as representing WENS (the most significant bit comes first, so N = 1, S = 2, E = 4, and W = 8) and these codes will give us NE, SE, SW and NW.

For more precise control, you can employ *half-step mode* using N, NE, E, SE, S, SW, W, and NW, taking eight half-steps for each electrical cycle. The corresponding numbers are 1, 5, 4, 6, 2, 10, 8, 9.

Try it!

3.2.2 Driving More Powerful Stepper Motors

Now we are ready to move on to the stepper motors that you will use for the trolley. These should have six wires, so that in addition to the N, S, E, and W connections, there are center taps to the two coils. Now these center taps can be connected to +12 V, and the drive circuit will pull one or two of the NSEW connections to ground.

A drive circuit that can do the job is a single-chip "octal Darlington driver" such as the ULN2803A, which can be connected directly to the output pins of the printer port. It has eight outputs that can be connected to the NSEW pins of two stepper motors. It even contains diodes to suppress the "spikes" when the inductive loads are turned off.

Connect the circuit as shown in Figure 3.1. Switch on and run the same software again, to make sure that the connections are in order. (When troubleshooting mechatronics, try to make only one change at a time, inching your way from one working system to the next.)

It is now time to test the second motor. Change a line of your `stepto` routine to

```
OUT port, n * 16
```

which will control the most significant 4 bits of the port.

The new motor should move in the same way as the other. Try

```
OUT port, n * 17
```

and both motors should move together.

3.2.3 The Mobile Robot

Mount the motors and wheels as shown in Figure 3.2, and you have the rudiments of a mobile robot. Now, however, we must write some much better structured software.

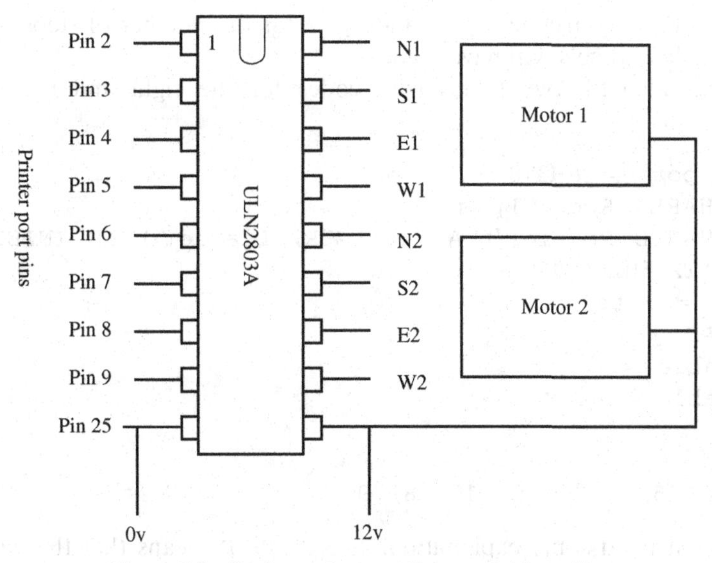

Figure 3.1 *Printer port connections and driver chip.*

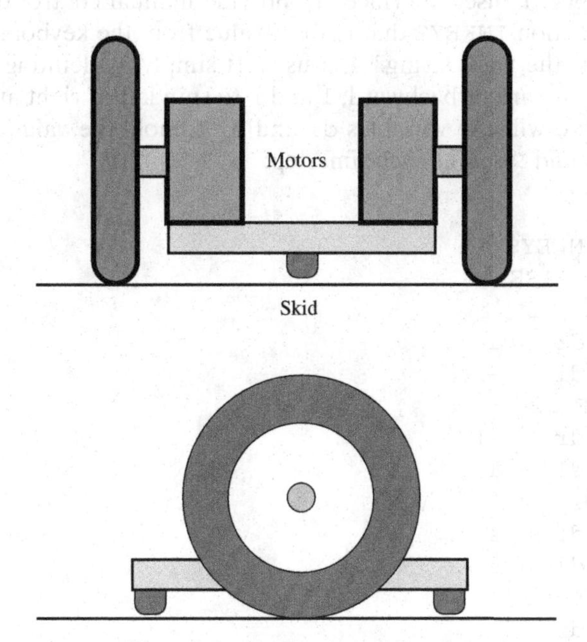

Figure 3.2 *Sketch of a mobile robot.*

Let us define variables Rpos and Lpos for the number of steps that the right and left wheels will have made.

Let us also fill two tables with codes for the right and left wheels, respectively:

```
CONST port = &H378
DIM SHARED Rpos, Lpos
DIM SHARED Rtable(7) AS INTEGER, Ltable(7) AS INTEGER
PLAY "T255L16"
FOR i = 0 to 7
   READ j
   Rtable = j
   Ltable = 16 * j
NEXT

DATA 1, 5, 4, 6, 2, 10, 8, 9
```

This might need some explanation. The SHARED means that the variables defined in the DIM statement will exist in all subroutines—otherwise subroutines can have their own private variables of the same name. READ picks up values one by one from a DATA statement.

Now we need a "user interface" to provide manual control of the robot. There is a function INKEY$ that grabs a value from the keyboard. If no key is pressed it is the "null string." Let us start simply by defining five keys, f and b to run forward or backward, l and r to spin left or right and the space bar to stop. We will use variables dl and dr to hold the value by which to change Lpos and Rpos at each timestep:

```
DO
   a$ = INKEY$
   SELECT CASE a$
      CASE "f"
         dr = 1
         dl = -1
      CASE "b
         dr = -1
         dl = 1
      CASE "l"
         dr = 1
         dl = 1
      CASE "r"
         dr = -1
         dl = -1
      CASE " "
         dr = 0
         dl = -0
```

```
    END SELECT
    Rpos = Rpos + dr
    Lpos = Lpos + dl
    DoMotors
LOOP UNTIL a$ = "q"
OUT port, 0
END
```

Note that when both wheels run forward, one motor must turn clockwise and the other one anticlockwise!

Now we need the `DoMotors SUB`. We can use the `MOD` operator to find `Rpos` modulo 8. That means that as `Rpos` increases, `Rpos MOD 8` cycles repeatedly through the numbers 0 to 7, just the thing we need to look up the drive value in the table:

```
SUB DoMotors
OUT port, Rtable(Rpos MOD 8)+Ltable(Lpos MOD 8)
PLAY "n0"
END SUB
```

Enter and run the program—do not forget to save it first. The trolley should obediently respond as you tap keys, stopping when you press the spacebar and the program ending when you press ⟨q⟩.

But that is just the start. From `Lpos` and `Rpos` you can calculate how far you have gone (you will have to include a variable representing the circumference of the wheels), and you can calculate your heading (you will have to include another variable to represent the separation of the wheels). Indeed, you can keep an estimate of your current position and change your program to accept target coordinates.

You can even add simple sensors. The printer port has input bits for "online," "out-of-paper," and several other conditions. You can read these bits with

```
inp(port+2)
```

You can include the following definition in your software:

```
pin15 = 8       'True if high,  used for error
pin13 = &H10    'True if high,  printer present
pin12 = &H20    'True if high,  out of paper
pin10 = &H40    'True if high,  -ack
pin11 = &H80    'True if low,   -busy
```

If you now add a contact that connects pin 12 to ground if the robot touches a wall, then

```
inp(port+2) AND pin12
```

will have the value zero if the wall is touched and 32 otherwise.

There are even some additional output bits at address `port+1`. These are not buffered and cannot drive a load, however. Be careful. Other bits on this port control the mode of the printer port, determining whether it is buffered or can act also as eight input bits.

```
pin1  = 1     'True gives low,  Strobe
pin14  = 2    'True gives low,  Auto linefeed
pin16  = 4    'True gives high, initialise
pin17  = 8    'True gives low,  select
```

3.3 BALL AND BEAM

In the mid-1960s the focus of my research was "Fast model predictive control for higher order systems." I needed a "higher-order system" on which to demonstrate its effectiveness and devised the ball-and-beam experiment, based on a childhood memory of a game in a seaside amusement arcade.

One outcome was that I discovered that it could be controlled just as efficiently by much simpler pragmatic methods. My research was a success, but my belief in the usefulness of "interesting" academic solutions was seriously undermined.

You might like to see an article linked at http://www.essmech.com/3/3.

3.3.1 Construction

A ball rolls in a grooved plank that is hinged at a central pivot. The motor is driven to tilt the plank, and the task is to control the position at which the ball comes to rest. In this stepper motor version, the ball will oscillate gently close to the target.

When manual control is provided as an alternative, it can be seen that the computer performs much better than a human being. The mechanical construction is clear from Figure 3.3.

There remains the problem of sensing the ball position. One variation is to mount a Webcam above the track and deduce the ball position from the "white blob" in the image. However, we will try a solution that uses much simpler technology.

The method of making the original sensor, and one that is still acceptable, is to stretch a resistance wire along the track. The steel ball makes contact between the wire and another wire on the opposite face of the "V," thereby forming a potentiometer.

The resistance of the wire is likely to be rather small, so a series resistor will have to be added to limit the current and avoid overheating. This means

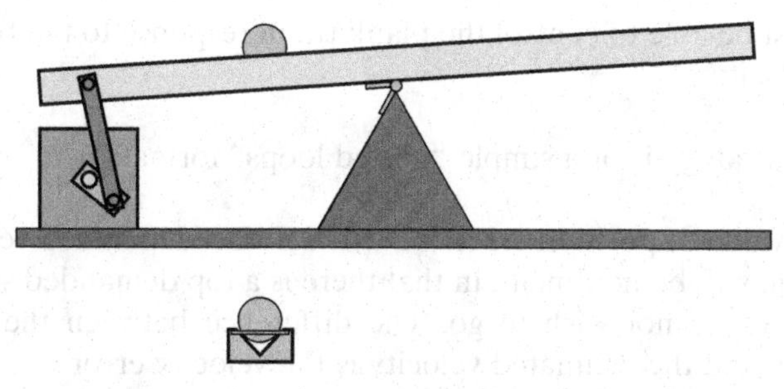

Figure 3.3 Mechanics of ball-and-beam experiment.

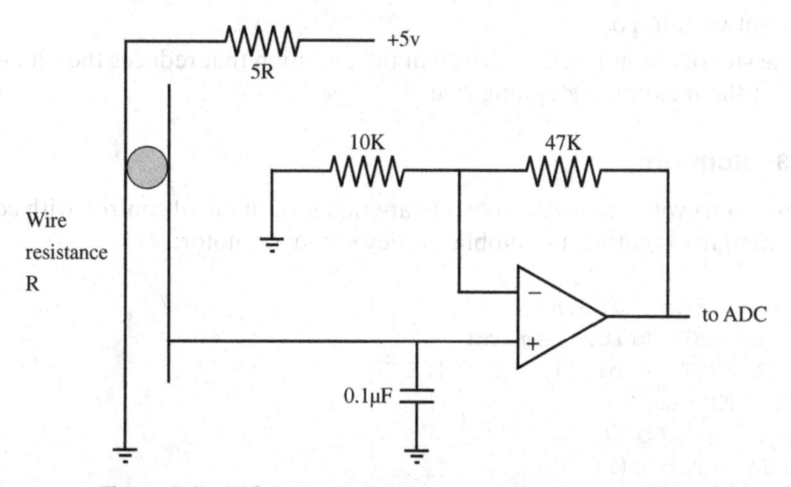

Figure 3.4 ADC input circuitry, including operational amplifier.

that the voltages will also be small and there may be a need to add an amplifier.

As the ball rolls, the contact will almost certainly break at times. This would cause the output voltage to suffer from steps to zero and would introduce a large noise signal. With the addition of a capacitor, maybe 0.1 μF (e.g., see Fig. 3.4), the output voltage stays constant when the circuit is broken and the noise consists simply of the step to the new value when contact is resumed.

3.3.2 A Control Strategy

Now we have several new points to establish before the task is complete:

- We must be able to read the ball position into the computer, by writing a routine to drive the ADC.
- We must be able to estimate the ball velocity.

- We must be able to control the plank tilt in response to the ball position and velocity.

The control strategy is of a simple "nested loops" format:

For each value of position error, we will define a demanded velocity. This function will be nonlinear, in that there is a top demanded speed above which we do not wish to go. The difference between the demanded velocity and the estimated velocity is the velocity error.

For each value of velocity error, we will demand a tilt angle. This function will be nonlinear, in that there is a maximum tilt beyond which we do not wish to go.

The stepper motor will be driven in the direction that reduces the tilt error at the maximum stepping rate.

3.3.3 Software

We will start with the version of software that gives manual control, with code very similar to that for the mobile trolley's stepper motor:

```
CONST port = &H378
DIM SHARED Tilt, Demand
DIM SHARED Table(7) AS INTEGER
PLAY "T255L64"
FOR i = 0 to 7
   READ Table(i)
NEXT

DATA 1, 5, 4, 6, 2, 10, 8, 9

DO
   a$ = INKEY$
   IF VAL(a$)>0 then
      Demand = (VAL(a$) - 5)/4  'a fraction -1 to 1
   END IF
   DoTilt
LOOP UNTIL a$ = "q"
OUT port, 0
END

SUB DoTilt
Tilt = Tilt + Demand
OUT port, Table(Tilt MOD 8)
PLAY "n0"
END SUB
```

You will need to use ⟨F1⟩ to discover what the function VAL does.

The number key that you hit will set the speed of the tilt motor, ⟨1⟩ tilting one way, ⟨9⟩ tilting the other way, and ⟨5⟩ causing it to stop.

3.3.4 The ADC Routine

This is a good time to add the ADC routine. It will differ for each interface technique, but the rest of the code can remain the same.

In Section 5.3.5, the code is given for interfacing a chip, the MCP3204, directly to the printer port. This is certainly the most economical and "future-proof" way to go if you do not already have an ADC card.

For most ADC cards, the principles are the same. The card has a base address, just as the printer port has address &H378. It will in fact use a range of 8 or maybe 16 addresses from the base upward. Writing to one of these addresses will set the channel number. Another will start the conversion.

Reading from one of the addresses will give access to a "busy" or "data ready" bit. The simplest code will enter a loop, repeatedly reading this bit until the answer is ready. It wastes a few microseconds, but is less prone to error than relying on interrupts.

If the converter is a 12-bit one, the lower 4 bits of the byte that contains the data-ready bit will contain the most significant 4 bits of the result. Yet another register will contain the lower 8 bits of the result, so we need only combine these and return to the program.

A little extra finesse is added if we scale the answer to lie in the range –1 to +1, so that the rest of the code is not changed if our precision is different. Here is an example. It is for a Contec Series 100 ADC12-16M board, so ancient that it will no longer fit into the PCI slot that is provided on current PCs:

```
CONST b = &H220                   'board base address
CONST ADlo = b + 4
CONST ADhi = b + 5
CONST chan = b + 10               'multiplexer channel
CONST start = b + 12
CONST busy = 16                   'busy bit in ADhi value

FUNCTION adc (c%)
DIM v%
OUT chan, c%                      'set channel number
OUT start, 255                    'tell ADC to start conversion
DO                                'wait until ready
   v% = INP(ADhi)
LOOP UNTIL (v% AND busy) = 0
v% = (v% AND 15) * 256 + INP(ADlo)
```

```
                                'combine high nibble
                                'with low byte
adc = (v% - 2048) / 4096  'change range to +/- 1
END FUNCTION
```

Here is the code for a different brand of board, an Advantech PCL-818L, also made to fit the older card slots:

```
CONST adstart = &H2A6
CONST adhi = &H2A7
CONST adlo = &H2A6
CONST readybit = &H40

FUNCTION adc(chan%)
DIM hibyte AS INTEGER
OUT adstart, &H80+chan%
DO
   hibyte = INP(adhi)
LOOP UNTIL hibyte AND readybit
adc = (INP(adlo) + 256*(hibyte AND 15) - 2048)/2048
END FUNCTION
```

A more recent Advantech card that does fit the newer slots is the PCI-1710. It provides the data in a 16-bit word, and to use it, it is necessary to extend QBasic with a library function that uses the INW function of the PC's micro. To add this library when you run, simply launch QBasic with the line `qbasic /linw`, either by launching it from the RUN control in the START bar or by making a shortcut on the desktop. You can edit the shortcut to add the extra `/linw`.

The ADC code and the library function have been supplied by Rodney Elliott of the University of Canterbury, New Zealand. You can find both at http://www.essmech.com/3/3/4.htm.

Two other alternatives are to use a "satellite" microcontroller attached to the serial port to perform the ADC conversion, or to construct the simple ramp-based converter described in Section 5.3.5.

So, having added the appropriate constants and ADC routine to our code, we can add some graphical output.

3.3.5 Graphics

After the constants have been defined, add

```
CONST Tmax = 4, dt=.012
SCREEN 12
WINDOW(-.1, -1.1)-(Tmax, 1.1)
LINE (0,0)-(Tmax,0),9
```

Now we plot the ball position by adding

```
t=t+dt
if t>Tmax then t=0
Ball = Adc(0)
PSET (t, Ball)
```

immediately after the `DoTilt` line.

This will give us a dot that crosses the screen in 4 seconds, since the note length that we have set is equivalent to 0.012 s.

You might need to make adjustments to the amplifier, if you use one, to get the value of the ball position to change over the range of –1 to 1. However, provided you have a reasonable range of change, you can replace the `Ball =` line with

```
Ball =  2 * (Adc(0) - Ballmin)/(Ballmax - Ballmin) -1
```

where `Ballmin` and `Ballmax` are the ADC values you have read at the two extremes of the plank. That will check out the hardware, but we still need to estimate the ball velocity.

Do you still remember the simulation of the lowpass filter in Section 3.1.3? It is no harder to simulate a highpass filter to make an estimate of the ball's speed. The whole thing reduces to two lines of code:

```
Ballvel = (Ball - Ballslow) * 10
Ballslow = Ballslow + Ballvel*dt
```

The argument is the same as before. `Ballslow` is a lowpass-filtered version of `Ball`, `just` as `x` was a low-pass-filtered version of `u` in Section 3.1.3. In this case, the *time-constant* smoothing the differentiation will be $\frac{1}{10}$s. Do not worry. There will be ample theory on this later in the book.

Add these lines, together with

```
PSET (t, Ballvel),14
```

after the existing `PSET` line in your code, and you should be able to see the movement of the ball and its velocity on your screen, as you command the plank to tilt.

3.3.6 The Strategy in Software

Now we just need to automate the process. But let us do it in stages.

First, let us make the demand input control the tilt target, rather than the tilt rate. Add a line at the top

```
DIM SHARED TiltDemand
```

and change the first line of code in DoTilt to

```
Tilt = Tilt + SGN(TiltDemand - Tilt)
```

(Use the ⟨F1⟩ key to see what SGN does.) Then add the line

```
TiltDemand = 20 * Demand
```

just before the DoTilt line in the main program, and you should find that the ball is a little easier to control. You might find that you need to change the number 20 to some other value that gives a useful range of tilt angles.

Of course, the computer does not "know" when the plank is level, so you must hold it level when you start to run the program.

For the next stage, let us make Demand control the velocity of the ball. Now we need something like

```
TiltDemand = kt * (0.2 * Demand - BallVel)
```

where kt is a *gain constant* that we would like to be large, to correct velocity errors quickly.

We do not want the tilt to be too great, because it takes time to drive the plank back to a level position. Let us define SUB Limit.

```
SUB Limit(x, lim)

IF x > lim THEN
   x = lim
ELSEIF x < -lim THEN
   x = -lim
ENDIF
```

Now the line

```
Limit TiltDemand, 20
```

after TiltDemand is calculated, will limit the tilt to 20 steps either way.

Make the necessary changes to your code, and experiment with values of the tilt gain, kt. Maybe you need to reduce the value of the tilt limit.

For the final step, make Demand set the target position of the ball.

In the final program, we will have the following constants and definitions at the top, to which must be added the constants needed for the ADC routine

```
CONST port = &H378
DIM SHARED Tilt, TiltDemand
```

```
DIM SHARED Table(7) AS INTEGER
PLAY "T255L64"
FOR i = 0 to 7
   READ Table(i)
NEXT

DATA 1, 5, 4, 6, 2, 10, 8, 9
CONST Tmax = 4, dt=.012
SCREEN 12
WINDOW(-.1, -1.1)-(Tmax, 1.1)
LINE (0,0)-(Tmax,0),9
```

while the main loop of our code will become

```
DO
   a$ = INKEY$
   IF VAL(a$)>0 then
      Demand = (VAL(a$) - 5)/4
   END IF
   Ball =  2 * (Adc(0) - Ballmin)/(Ballmax - Ballmin) -1
   BallVel= (Ball - Ballslow) * 10
   Ballslow = Ballslow + BallVel * dt
   t = t + dt
   IF t > Tmax then t = 0
   PSET (t, Ball)
   PSET (t, BallVel), 14
   VelDemand = kv * (Demand - Ball)
   Limit VelDemand, Velmax
   TiltDemand = kt * (VelDemand - BallVel)
   Limit TiltDemand, TiltMax
   DoTilt
LOOP UNTIL a$ = "q"
OUT port, 0
END
```

You will have to add lines at the top to set the values that you choose for `kt`, `kv`, `VelMax` and `TiltMax`. You might also wish to change the differentiator time constant, which is at present set to 10.

As well as the code above, you will have an appropriate ADC routine and

```
SUB DoTilt
Tilt = Tilt + SGN(TiltDemand - Tilt)
OUT port, Table(Tilt MOD 8)
```

```
PLAY "n0"
END SUB

SUB Limit(x, lim)
IF x > lim THEN
   x = lim
ELSEIF x < -lim THEN
   x = -lim
ENDIF
```

You still have to hold the plank level when the program starts to run.

3.3.7 The Next Step

Now that your system has sensor inputs, the system can discover many of its own parameters. The demonstration version can include switching between control modes and can perform its own calibration at startup. This procedure is as follows:

1. Tilt the plank left 100 steps. This will ensure that the plank hits its limit stop, after which the motor will just "shudder" as more steps are commanded.
2. Wait 3 s for the ball to run down the plank. Measure the value of `Adc(0)` and store it as `BallMin`.
3. Tilt the plank slowly to the right, until the value of `Adc(0)` starts to change. Here the plank will be level, so we would like to set `Tilt` to zero. However, we want to output that particular NSEW combination to the stepper motor to level the plank. So first we set a variable `Tilt0` to `(Tilt MOD 8)`, then set `Tilt` to zero and afterward use (`(Tilt0 + Tilt) MOD 8`) for looking up values from the table.
4. Tilt the plank to `Tilt=20` and wait 3 s, or until the ball stops moving.
5. Measure the value of `Adc(0)` and save it as `BallMax`.

Now you can enter the loop for ball position control.

3.4 "PROFESSIONAL" POSITION CONTROL

An axis of an industrial robot is a far cry from the usual laboratory position control experiment. Most students will be content to derive a response curve that matches the classical "damped second-order response" found in books on linear systems. The professional designer of a motion controller will reject this out of hand. He or she will expect the system to stop at the target position as though hitting a brick wall and to resist all deflecting forces with only the slightest perturbation.

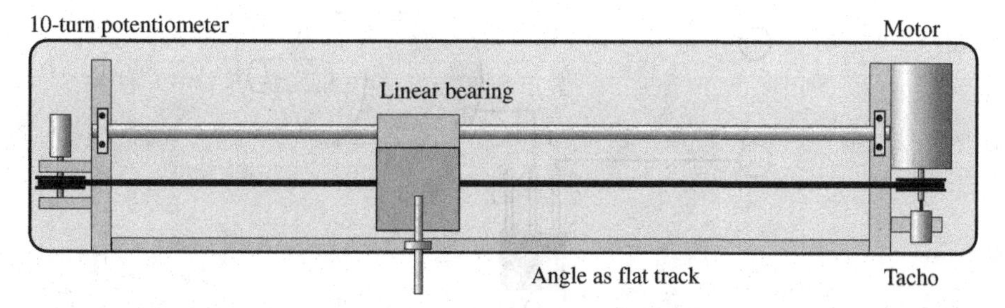

Figure 3.5 *Position control hardware.*

Far too many laboratory experiments are protected by a transparent cover. This prevents the student from feeling the stiffness of the output. The present experiment is meant to be poked and prodded unmercifully.

The hardware (see, e.g., layout in Fig. 3.5) forms three-quarters of an inverted-pendulum experiment that follows, although the control strategy is quite different.

There are many ways to construct this experiment. The machine in our mechatronics laboratory has survived several generations of students, although the new version illustrated here has just been completed. The original motor was marketed as a component for a battery lawnmower, but many suitable motors are now sold for adding electric drive to bicycles.

The 100 W DC motor drives a pulley at one end of a toothed belt. The belt pulls a trolley weighing half a kilogram along a track, constrained on one side by a linear bearing and on the other by a ball race running on a flat track formed by a length of angle.

The belt is joined to the trolley on one side, passes around the motor pulley, back under the trolley, around a second pulley at the other end of the track, and back to the other side of the trolley. The second pulley turns a 10-turn potentiometer. Also connected to the motor shaft is a small motor that acts as the tachometer.

Care must be taken to avoid the rotating parts binding, when bearings are aligned. The potentiometer and the tacho are mounted in sprung clips, so that they can float to accommodate any misalignment.

Two separate power supplies should be used. The sensors and ADC circuitry will use one supply while a second power supply drives the motor. If sensors and motor were to share the same supply, there would be a risk of high-frequency oscillation as the motor drive affected the sensor supply voltage. Another advantage of separating them is that the motor supply can be kept at zero when the program starts, increasing toward 12 V only as the student's full attention is on any oscillations or excursions of the system.

In the original experiment, the potentiometer is supplied by +15 and –15 V supply lines, so that zero volts will represent the center of the track. In the new version, using the single-chip converter, it is supplied from 0 and 5 V,

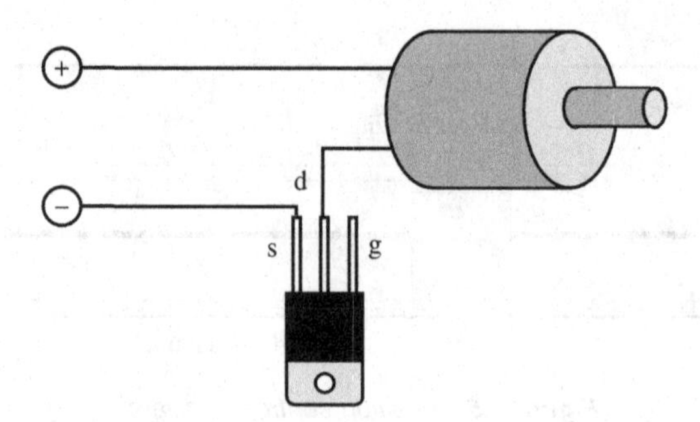

Figure 3.6 Sketch of motor, FET, supply.

with the ADC software arranged so that zero is read in the middle of the range. In this case the tacho "ground" should be connected to the midpoint of two 1 kΩ resistors connected across the 5-V supply.

There are a number of steps that we need to accomplish on the way. First we must control a motor of up to 100 W, driving it bidirectionally. We must also consider the use of mark–space control, so that the motor is not continuously under full power.

We also look at the effect of "tacho feedback" for reducing the effect of disturbances.

3.4.1 Simple Motor Control

The experiment can start with a variable DC power supply, a small DC motor, and an N-channel power field effect transistor (see Fig. 3.6):

1. Connect the motor to the power supply. Increase the voltage—probably to around 10 V—so that the motor starts readily and runs at a reasonable speed.
2. Switch off and disconnect the negative connection. Connect the field effect transistor in series with the motor. The *source* is connected to the negative lead of the power supply, while the motor is connected to the *drain*. The *gate* is left unconnected.
3. Switch on again. Hold the negative power supply contact and touch the gate with your finger. The motor will not run. Hold the positive lead instead and touch the gate. The motor should run as it did before connecting the transistor.

A suitable power transistor for this part is a BUK553 N-channel FET. It is designed to be controlled by TTL (transistor–transistor logic) voltages, so you can connect a computer output line directly to the gate.

This next part of the experiment again uses the parallel printer port of the computer. The pins of interest are 2 to 9 for the output bits and pin 25 to serve as a ground pin. As before, the port address will either be `&H278` or `&H378`.

Run QBASIC.EXE, then press the function key ⟨F6⟩ to select the IMMEDIATE WINDOW. This allows you to run code at once, rather than saving it in a program. Enter the line

```
OUT &H378, 255
```

followed by RETURN. You should use a meter or an oscilloscope to check the voltage of pin 2—it should be at 5 V. In fact, this command should have set all eight lines from pin 2 to pin 9 to 5 V.

Now enter

```
OUT &H378,0
```

followed by RETURN. The voltage should fall to zero.

If these actions do not work, repeat them using `&H278` instead. In that case also use 278 instead of 378 in the program below. Enter the following program to make changing the outputs neater and more convenient:

```
port = &H378

DO
   a$=INKEY$
   IF a$<>"" THEN
      OUT port, VAL(a$)
   END IF
LOOP UNTIL a$="q"

OUT port, 0
```

If you press the ⟨1⟩ key, pin 2 should be at 5 V and pin 3 at zero. Press the ⟨2⟩ key, and pin 3 should be at 5 V and pin 2 at zero. Press the ⟨0⟩ key, and both should become zero. Press ⟨q⟩ and the program should end, tidying up by setting the port to zero. Save it as P2.BAS.

Now switch off the motor power supply. Connect the computer ground (pin 25) to the negative line and pin 2 to the gate of the transistor. Switch on again. By tapping the ⟨1⟩ and the ⟨0⟩ keys, you should be able to start and stop the motor at will.

Of course, the motor will also run if you tap ⟨3⟩, ⟨5⟩, ⟨7⟩ or ⟨9⟩ and stop if you choose an even number.

3.4.2 Unidirectional Speed Control

The "safe" version of hardware for this part of the experiment uses two miniature DC motors (see Fig. 3.7) that require only a few watts to drive them. Later the much larger motor of the position control experiment can be used,

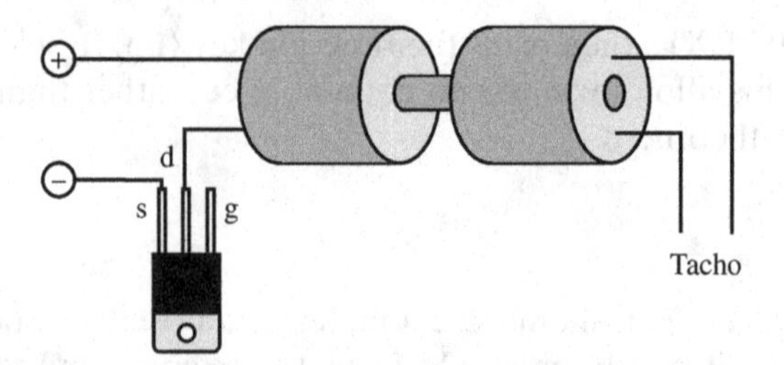

Figure 3.7 *Two small motors with FET.*

with the belt disconnected, so that the trolley does not move. This motor uses much greater power, and care must be taken to avoid overheating the field effect transistors used in the experiment.

You should first run through these experiments with the miniature motors, then repeat them with the larger motor. In each case the driven motor is connected mechanically to a second DC motor. Both turn together. When they run, a voltage is generated on the second motor that is proportional to the speed. By connecting this voltage to an analog input of the computer, you can use the speed as a feedback signal. This second motor is being used as a tachometer.

As in previous experiments, the ADC routine will first read in an integer and then scale it to a floating-point value in the range –1 to 1. You need to find the code that is appropriate for your ADC and enter both the FUNCTION and the constants that define the interface.

With this ADC routine and with the velocity voltage connected to channel 0, the start of a program to view the tacho signal can be as follows:

```
CONST tmax=4
SCREEN 12

WINDOW(0,-1)-(tmax,1)

dt = .01

DO

   a$ = INKEY$
   v = ADC(0)
   t = t + dt
   IF t > tmax THEN t = 0
   PSET (t, v)
LOOP UNTIL a$ = "q"
```

This will read the voltage and display it on the screen as a sort of *oscilloscope display.* Spin the motors by hand, then run the driven motor by "dabbing" a wire across source and drain of the control transistor.

One of the most important tasks in feedback control is to ensure that the signals are of the correct polarity. If they are not, negative feedback can become positive feedback with disastrous results.

When you run the motor, make sure that the resulting velocity is shown on the screen as positive. If it is not, reverse the connections of the tacho.

Pressing the ⟨q⟩ key will end the program.

When you are sure that this part of the program is working correctly, you can expand it to obtain closed-loop control. You need to include a command for turning the motor on and off and another for comparing the speed with some demanded value. In turn, something is needed for changing the demand value at the press of a key:

```
CONST port= &H378
CONST tmax=4

SCREEN 12
WINDOW (0,-1) - (tmax,1)

dt = .01

vdemand = 0

DO
   a$ = INKEY$
   if a$ = ">" THEN vdemand = vdemand + 0.1
   if a$ = "<" THEN vdemand = vdemand - 0.1
   v = ADC(0)
   IF vdemand > v THEN OUT port, 1
   IF vdemand < v THEN OUT port, 0
   t = t + dt
   IF t > tmax THEN t = 0

   PSET (t, v)
   PSET (t, vdemand), 12 'red

LOOP UNTIL a$ = "q"

OUT port, 0
```

Run the program. The motor should remain still. Now tap the ">" key once or twice (remember to press ⟨shift⟩), and the motor should turn. Hold the motor shaft loosely to try to slow it down. You should see the motor current increase as indicated on the power supply meter, with very little drop in speed.

In practice it is preferable to use "." in place of ">" and "," in place of "<" so that there is no need to press the ⟨shift⟩ key.

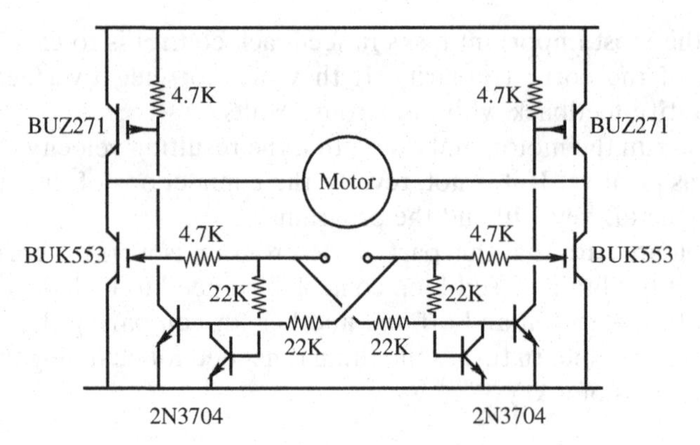

Figure 3.8 *H-bridge schematic.*

3.4.3 Bidirectional Speed Control

When the motor turns in just one direction, the speed can be controlled simply by turning the motor on and off. If the motor is to be capable of turning both ways, we must have some means of driving it with either a positive or negative voltage. We could use two power supplies to give positive and negative voltages, or we can use a single power supply with an *H-bridge* (see Fig. 3.8).

As we have seen, this enables each wire of the motor to be connected either to the positive supply or to ground. The schematic has the appearance of the letter "H," giving it its name. The circuitry must at all costs prevent one side being connected to both positive and ground at the same time!

The circuit uses two bits of the parallel printer port, bit 0 and bit 1 on pins 2 and 3. When the output value is 1, the motor runs in one direction; when it is 2, it runs in the opposite direction; and if it is zero, the motor free-wheels until it stops. (Remember that the binary value of bit 0 is 1; of bit 1, is 2; of bit 2, is 4; and so on.)

Start off by using the program of the last section. You will have changed the connections to the motor in order to replace the single transistor by a full H-bridge drive. In the process, the sense of the motor drive could have changed.

When you tap the ">" key, the motor should start to run. If the velocity trace on the screen is positive, all is well. If not, you must reverse the motor leads or the tacho leads. But which?

If you are using the position control rig with the belt removed, check that the motor runs in a sense that would carry the trolley to the right. If not, reverse it. Now check the tacho voltage and if necessary reverse the tacho connections to ensure that the trace is positive.

You should now get the same velocity control as before—simply running in the positive direction. Now is the time to change the software to take advantage of the two-way drive.

The previous program needs only slight modification to give "bang-bang" control. If the velocity error (v – vdemand) is positive, full negative drive is applied. If the error is negative, the drive is positive.

```
CONST port= &H378
CONST tmax=4

SCREEN 12

WINDOW (0,-1) - (tmax,1)
dt = .01

vdemand = 0

DO
   a$ = INKEY$
   IF a$ = "." THEN vdemand = vdemand + 0.1
   IF a$ = "," THEN vdemand = vdemand - 0.1
   v = ADC(0)
   olddrive=drive
   IF vdemand > v THEN
      drive = 1
   ELSEIF vdemand < v THEN
      drive = 2
   ELSE
      drive = 0
   END IF
   IF drive = olddrive THEN
      OUT port, drive
   ELSE
      OUT port, 0
   END IF
   t = t + dt
   IF t > tmax THEN t = 0
   PSET (t, v)
   PSET (t, vdemand), 12 'red
LOOP UNTIL a$ = "q"
OUT port, 0
```

So, why is the code complicated with `drive` and `olddrive`? If we switch repeatedly from forward to reverse drives, there is a tendency for the H-bridge transistors to overheat. Here a zero is output between changes of sign.

Nevertheless, there is the disadvantage that for most of the time maximum drive is applied. Is there a way to reduce the drive when full drive is not needed?

3.4.4 The Proportional Band

The previous strategy resulted in the motor being connected to the power supply at all times, resulting in a substantial power drain even when there is no disturbing torque and the demanded speed is zero.

The current drain could be reduced if the motor were switched off when "near" the target. There would, in effect, be a "gap" where there is zero drive.

Add a line to the program above:

```
CONST gap = .05
```

Then change the "middle" lines to

```
IF vdemand > v + gap THEN
   drive = 1
ELSEIF vdemand < v - gap THEN
   drive = 2
ELSE
   drive = 0
END IF
```

This will reduce the current drain—indeed, it will be zero when zero speed is demanded—but the velocity can be in error by an amount "gap" with no corrective action. With an on off controller, how can we get proportional action to fill in the gap? Many commercial power amplifiers provide a proportional mark-space output—at great expense. We can construct a mark–space controller in software.

We set up a variable `g` that will shuttle to and fro across the gap in a triangular wave, as follows:

```
g = g + dg
IF g > gap THEN dg = -.1 * gap
if g <= 0 THEN dg = .1 * gap
```

Now we change the "engine room" lines again, to

```
IF vdemand > v + g THEN
   drive = 1
ELSEIF vdemand < v + g - gap THEN
   drive = 2
ELSE
   drive = 0
END IF
```

dg must be set to 0.1 * gap at the top of the program, after gap is defined. After 20 times round the loop, g cycles through its range of values. If the velocity error is exactly zero, the drive will be set to zero all the time. If vdemand - v = gap/2, the drive will be positive for half the time. If it is greater than gap, the drive will be positive all the time.

We have a mark–space ratio that can increase the drive in steps of 10%, which is dg's proportion of gap. If dg is made much smaller, the cycle time is longer and the motor buzzes accordingly.

3.4.5 Position Control

By now you will have applied velocity control to the motor and tacho of the position control experiment, having carefully removed the belt from contact with the motor pulley.

Instead of using the keyboard to set the velocity demand, we can use the 10-turn potentiometer mounted on the second pulley.

Connect the potentiometer output to the input of ADC channel 1. The software function should now return a value that varies from –1 to 1 as the potentiometer is turned from end to end:

```
CONST port= &H378
CONST tmax=4
CONST gap=.05

dg = .1 * gap
k = 1

SCREEN 12
WINDOW (0,-1) - (tmax,1)

dt = .001 'or a smaller value to suit the display rate
DO
   a$ = INKEY$
   v = 5 * ADC(0) 'The tacho voltage is rather small
   x =ADC(1)
   vdemand = -k * x
   g = g + dg
   IF g > gap THEN dg = -.1 * gap
   IF g <=0 THEN dg = .1 * gap
   verror = vdemand - v 'more about this later
   IF verror > g THEN
      OUT port, 1
   ELSEIF verror < g -gap THEN
      OUT port, 2
```

```
    ELSE
       OUT port, 0
    END IF
    t = t + dt
    IF t > tmax THEN t = 0
    PSET (t, v)
    PSET (t, x), 14 'yellow
LOOP UNTIL a$ = "q"
OUT port, 0
```

As a consequence of the way we have used g and gap, we should always have one or more periods of zero drive before a drive reversal, so we can dispense with drive and olddrive.

Run the program. The potentiometer will control the speed. If the polarities are correct, the motor should spin in the opposite sense to the way you turn the potentiometer shaft. If this is not the case, reverse the supply connections to the potentiometer.

Now switch off the power and reconnect the drive belt. Beforehand, be sure that the speed control is "good".

Cautiously increase the voltage of the power supply that runs the motor. As the motor starts to move, it backs off the potentiometer shaft to zero—you have achieved position control.

The next step is to put back a demand signal, this time a position demand:

```
if a$ = "." THEN demand = demand + 0.1
if a$ = "," THEN demand = demand - 0.1
if a$ = "0" THEN demand = 0
```

The vdemand line is now

```
vdemand = k * (demand - x)
```

Tapping a key will step the demand along the potentiometer travel. Pressing the ⟨0⟩ key will return the demand to zero, so that a larger step response can be seen.

Experiment with various values of k and also gap, trying to get a fast response without overshoot.

Now put the control system to its real test. How far can the trolley's position be pushed away from the target before full corrective drive is applied? With linear feedback tuned to avoid overshoot, the control might be rather "soggy." With a nonlinear strategy, something much "crisper" can be achieved.

Try setting gap to zero—but make the test very brief, since the power transistors will be getting hot. This reverts to bang-bang control, and the system will be very stiff indeed, although the motor will buzz and the current will be high. Increasing `gap` will make the performance "quieter" at the expense of a softer response to disturbances.

3.4.6 Nonlinear Correction

An overshoot occurs if the motor approaches the target "too fast to stop." Drive saturation plays an important part in the performance. Even when the linear parameters have been tuned for a heavily damped response for small deflections, a larger disturbance can cause it to overshoot badly.

An easy answer is to put a limit on the demanded velocity. Now the system will approach the target at constant speed, however large the deflection. By setting this speed limit lower, the time constant of the final settling response can be made faster while still avoiding an overshoot. There is always a compromise to be made.

Add the line at the top of the program

```
vmax = 0.1
```

and after `vdemand` is defined, add the lines

```
IF vdemand > vmax THEN vdemand = vmax
IF vdemand < - vmax THEN vdemand = -vmax
```

Now experiment by varying the values of `k`, `vmax`, and `gap`.

3.4.7 Estimating Velocity

Suppose that we have no tacho signal. How can we stabilize the system?

In the ball-and-beam experiment, you produced an estimate of the ball's velocity in real time. The play routine was used to control a precise time increment, a necessary part of the process.

We used the property that a highpass filter can be expressed as the difference between the original signal and the output of a lowpass filter as follows:

$$Ts/(1+Ts)=1-1/(1+Ts)$$

In other words, to estimate the velocity, we construct a "lagged" version of the position and subtract it from the position. We can set up a "chicken and egg" situation where we use `vest` to update the lagged version and in turn calculate `vest` from the lagged version

```
vest = (x - xslow) * kt
```

while the lagged position is updated by

```
xslow = xslow + vest * dt
```

If the variables are updated at intervals `dt`, the time constant of the lag will be `1/kt` seconds, so if `kt` = 20, it is 50 ms.

Add a line at the top of the program:

```
CONST kt = 20
```

Now the scale factor of `vest` is unity, so that a value of 1 indicates that `x` is changing at 1 unit per second. In contrast, `v` was scaled by an arbitrary factor determined by the tacho.

The gain and the value of `vmax` you must use with `vest` will differ considerably from the values you chose when using `v`. You will find that `vest` would go off screen if displayed at the same scale as `v`; so display `vest/20` instead.

Our previous way of getting a timed response was by inserting a `PLAY` command into the program loop, slowing down the whole process. We would instead like to run the mark–space calculation as fast a possible. QBasic allows us to set up an "interrupt" so that `vest` will be updated in time "stolen" from the main loop every 10 msec.

Since the earliest days of computers, interrupts have been a fundamental part of the system. When your output device was a teletype, tapping away at 10 characters per second, you did not want to waste time waiting for it and instead preferred to get some more computing done between taps.

So, every time the interface is ready for another character, it interrupts the computer. The values in the registers used for the task in hand are tucked away safely, then the machine attends to loading the code of the next character to be printed and outputting it. Then it must retrieve and restore the values of the registers and perform an "interrupt return." Modern peripherals may be much faster, but they still operate at a snail's pace compared with computing speeds.

In just the same way, QBasic allows us to play music in the background, interrupting us for some more notes when the tune is coming to an end. If we set the play rate to the highest speed and supply just one note at a time, we will receive interrupts at intervals of 0.012 s.

Near the top of the program, the command

```
PLAY "mbl64t255"
```

sets the fastest playing speed and also tells the music to play in "background mode."

Add some further instructions just before the loop begins:

```
ON PLAY(1) GOSUB rates
PLAY ON

PLAY "cde"
```

These tell the software that when the number of notes queued for playing falls to 1, there should be an interrupt causing a subroutine call to the label `rates`. The third line "sets the alarm clock" with three notes to play.

Now at the end of the program, add a line

```
END
```

so that program execution cannot "fall through" to the subroutine.

Now for the subroutine itself. This is added after the `END`:

```
rates:
vest = (x - xslow) * kt
xslow = xslow + vest * dt
PLAY "n0"
RETURN
```

Add one last line in the heart of the control loop to display the estimated velocity in red, among the other `PSET` lines:

```
PSET (t, vest / 20), 12
```

Now run the program again. Remember that the actual control program is exactly as before—we are using "real" velocity and not `vest`.

As you demand steps of movement, the tacho velocity is shown in white. The estimated velocity is shown in red. If you have "got it right," the values in the constant speed section will be the same size. Try values other than 20 to scale `vest` to match the traces.

When you are at last satisfied with your estimate of the velocity, you can try it in the control loop. In the `verror =` line, replace `v` with `vest/20` (or the value you found for the best match). Now your control depends on estimated velocity, not on the tacho.

Control will be a bit more wobbly, overshoots will be harder to avoid, and the control may have to be softer. Once more, experiment with `k`, `vmax`, and `kt` to see what you can achieve.

3.4.8 Discrete-Time Control

There is just one more step to try. We will move the entire control loop into the interrupt routine. Now the top-level program merely "twiddles its thumbs" in a loop. Every 10 ms, it is interrupted to allow the ADCs to be measured, the feedback to be calculated, and the drive to be output.

The mark–space drive behavior given by `gap` would become a nuisance if `g` were changed only every 10 ms. Its cycle through 20 steps would take 200 ms. The system would vibrate at five cycles per second.

We can instead calculate a drive signal `u`

```
u = (vdemand - v) * kv
```

inside the interrupt routine, but let the mark–space gap part of the routine run much faster in what is left of the main program. With a value of 1.0 for gap, the drive will be positive all the time if u is greater than 1, will be negative all the time if u is less than –1, and will give a proportional mark–space ratio in between. The result is like this:

```
'CONSTants required for the ADC routine go here

CONST port= &H378         '(might be &H278)
CONST tmax= 4
CONST gap= 1
dg = .1 * gap
CONST vmax = 0.1

k = 1
kv = 10
kt = 10

SCREEN 12
WINDOW (0,-1) - (tmax,1)

PLAY "mbl64t255"
dt = .01                  'Interval for this playing rate

ON PLAY(1) GOSUB rates
PLAY ON
PLAY "cde"

DO
   a$ = INKEY$
   if a$ = "." THEN xdemand = xdemand + 0.1
   if a$ = "," THEN xdemand = xdemand - 0.1
   if a$ = "0" THEN xdemand = 0
   g = g + dg
   IF g > gap THEN dg = -gap/10
   if g < 0 THEN dg = gap/10
   IF u > g THEN
      OUT port, 1
   ELSEIF u < g - gap THEN
      OUT port, 2
   ELSE
      OUT port, 0
   END IF
```

```
LOOP UNTIL a$ = "q"
PLAY OFF
OUT port, 0
END

rates:
v = ADC(0)
x = ADC(1)
vest = (x - xlsow) * kt
xslow = xslow + vest * dt
vdemand = k * (xdemand - x)
IF vdemand > vmax THEN vdemand = vmax
IF vdemand < - vmax THEN vdemand = -vmax
u = (vdemand - v) * kv '(or later try vest)
t = t + dt
IF t > tmax THEN t = 0
PSET (t, v)
PSET (t, x), 14          'yellow
PSET (t, vest / 20), 12 'red

PLAY "n0"
RETURN

FUNCTION adc(chan%)

' The code for the ADC routine goes here

END FUNCTION
```

Experiment with various values of k, kv, and vmax.

Now try using vest instead of v for the control by changing the line suggested.

3.4.9 Summary

Now you have seen how a motor can be switched by one computer output line if it runs in just one direction, or by two lines with the aid of an H-bridge if it is to be bidirectional.

You have seen that a tacho velocity sensor enables the position control to be very stiff, as would be required by a machine tool positioner. You have seen that without such a tacho, a softer control is achievable with the use of a digital filter to estimate the velocity.

You have seen that the gain can be expressed in terms of a *proportional band* and that this proportional drive can be achieved as mark–space modulation by means of further software.

In short, you have seen that a motor, an amplifier, and two transducers can be turned into an *industrial-grade* position control system with a few simple lines of software.

3.5 AN INVERTED PENDULUM

In the previous experiment, a position control loop has been closed after some cautious tests. A variety of nonlinear strategies have been investigated to obtain the performance expected of an actuator such as a robot axis.

Most of the feedback decisions were made empirically, experimenting to find values that would give a swift response with no overshoot. For the control of an inverted pendulum, we can also follow an empirical approach. However, we must avoid the mistake of thinking that we can add the pendulum control on top of the algorithm that we have found for position control. You may find some of the feedback coefficients surprising.

3.5.1 Skeleton Software

We can strip out the control algorithm from the position control software, to leave the following skeleton:

```
'Constants required for the ADC routine go here

CONST port= &H378          '(might be 278)
CONST tmax= 4
CONST gap= 1
dg = .1 * gap
CONST vmax = 0.1

k = 1
kv = 10
kt = 10
SCREEN 12
WINDOW (0,-1) - (tmax,1)

PLAY "mbl64t255"
dt = .01              'Interval for this playing rate

ON PLAY(1) GOSUB rates
PLAY ON
PLAY "cde"

DO
   a$ = INKEY$
   if a$ = "." THEN xdemand = xdemand + 0.1
```

```
   if a$ = "," THEN xdemand = xdemand - 0.1
   if a$ = "0" THEN xdemand = 0
   g = g + dg
   IF g > gap THEN dg = -gap/10
   if g < 0 THEN dg = gap/10
   IF u > g THEN
      OUT port, 1
   ELSEIF u < g - gap THEN
      OUT port, 2
   ELSE
      OUT port, 0
   END IF
LOOP UNTIL a$ = "q"
PLAY OFF
OUT port, 0
END

rates:
v = ADC(0)
x = ADC(1)
vest = (x - xlsow) * kt
xslow = xslow + vest * dt
'The control goes here, with a line u = . . .

t = t + dt
IF t > tmax THEN t = 0
PSET (t, v)                'white
PSET (t, x), 14            'position in yellow

PLAY "n0"
RETURN

FUNCTION adc(chan%)

' The code for the ADC routine goes here

END FUNCTION
```

3.5.2 The Pendulum and Tilt Sensor

You might have noticed the tubular rod projecting from the front of the new position control experiment. This is a tube in which a pair of crossed Hall effect sensors are mounted, forming the pivot for the pendulum. Simple rubber O-rings restrain the pendulum mounting and prevent it dropping off.

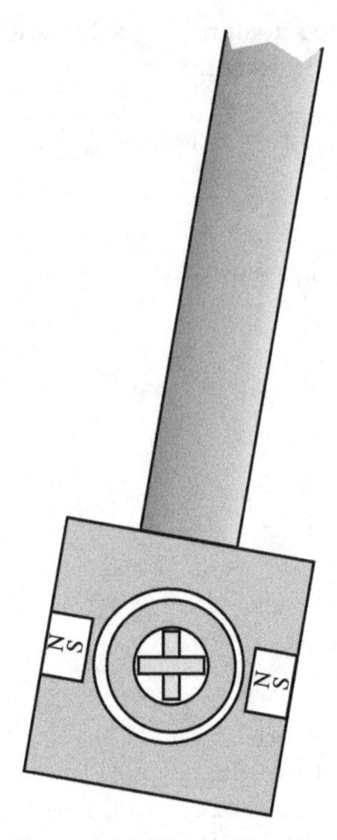

Figure 3.9 *Pivot and magnets.*

The pivot of the pendulum takes the form of a mounting for a pair of magnets, as shown in Figure 3.9. This slides onto the sensor tube. Into it can be screwed a variety of lengths of lightweight aluminum tubing.

The sensors are linear Hall effect sensors, chips the size of a small transistor with three connections. The UGN3504 is now obsolete, but equivalents such as the A1302 are available from Allegro. Supply lines of 5 and 0 V are connected to two of these, while the third delivers an output voltage that varies either side of 2.5 V in proportion to the normal component of the magnetic field.

With two sensors, both sine and cosine signals are available. For calculation and simple control, we can take the sine to be the same as the angle in radians over the small relevant range. However, we also have the information we need it we wish to swing the pendulum from hanging down to standing up.

The new version of the experiment uses the single-chip ADC described in detail in Section 5.3.4. Both input and output are handled by the printer port connection, while the signals from the Hall effect sensors are connected to inputs ADC2 and ADC3, pins 3 and 4 of the MCP3204 chip (see schematic in Fig. 3.10).

Add the following lines in the skeleton program above, in the subroutine `rates`:

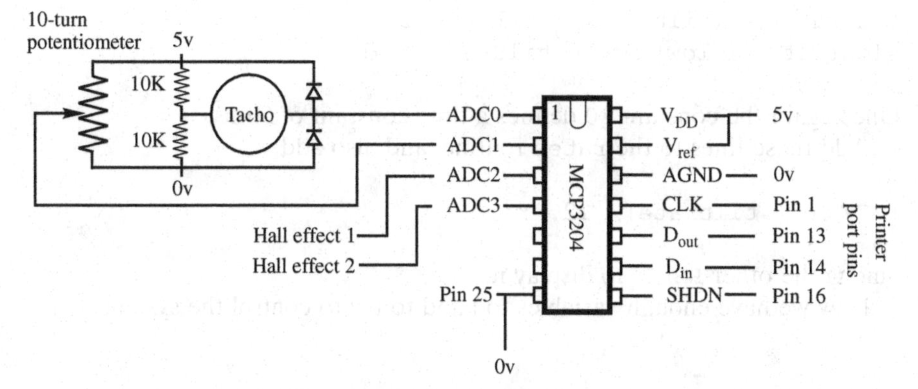

Figure 3.10 Circuit and printer port connections.

```
tilt = ADC(2)
PSET (t, tilt), 10 'Plot tilt in green
```

Make sure that the motor drive supply is switched off, and run the program. Rotate the sensor—swing the pendulum about—and note that the sensor voltage (shown in green) varies over a reasonable range. You may need to swap the sensor connections, so that `ADC(2)` represents the sensor that gives a value near zero when the pendulum is upright. The signal should move positive as the pendulum is tilted to the right. If it does not, slide the mounting off the pivot, reverse it front to back and replace it.

To centralize the tilt reading, we have to subtract the value given when the pendulum is straight up. Change the `tilt=` line to

```
tilt = (ADC(2) - tilt0)
```

But how do we get the value of `tilt0`? After the `dt=.01` line at the top of the code, add

```
tilt0 = adc(2)
```

and make sure that you balance the pendulum before running the program.

Rotate the pendulum and check that the value swings equally positive and negative.

3.5.3 Finding the Tilt Rate

Now we use the interrupt routine again to estimate the tilt rate from the tilt signal. Since we are using a substantial gain already, this estimate might be rather noisy:

```
tiltrate = (tilt - slowtilt) * 20
slowtilt = slowtilt + tiltrate * dt
```

Once again the constant 20 defines a time constant of 50 ms.

Add these lines to the rates routine, and also add

```
PSET (t, tiltrate), 12 'red
```

among the other PSETs to display it.

Now we have enough variables to hand to try to control the system.

3.5.4 Building a Strategy

Start with the skeleton program with the modifications of the last section. Before going any further, we should make sure that the polarities are correct. Run the program, but be sure to keep the motor power supply turned off.

Move the trolley to the right by hand, and make sure that the yellow trace rises on the screen. Move it swiftly a little way to the right and make sure that the white velocity trace also rises. Lean the pendulum to the right—the green trace should rise. Hold the pendulum straight up, and the green trace should be central. Wave the pendulum to the right, and the red tiltrate trace should be positive for a short burst.

If all is well, we are ready to begin.

The whole feedback exercise comes down to finding a suitable expression for the u = line. We can start with

```
u = 10 * tilt
```

If the pendulum tilts to the right, the trolley moves to the right. This implies that the trolley will move to try to hold the pendulum upright.

Hold the pendulum high up, with the trolley in the center of the track. Being careful to keep your hands clear of the trolley, turn the motor power supply voltage setting down to zero, switch on, and increase the volts steadily to 10 or so.

You will see that this feedback arrangement acts as a sort of position control. As you move the top of the pendulum to the left, the trolley moves to follow it. There is no damping, however, so the response will be fairly oscillatory. We should add some damping. Do this by changing the vital line to

```
u = 10 * tilt + 10 * tiltrate
```

By adjusting the two coefficients, you can obtain a swift and agile response. Try releasing the pendulum briefly—it should remain upright, but the trolley will drift to the left or right until the pendulum hits the stops.

It is time to add feedback to keep the trolley in the center of the track. But should we add positive or negative position feedback? Change the line again to

```
u = a * tilt + b * tiltrate + 10 * x
```

where a and b are the numbers you have chosen by trial and error.

Hold the pendulum again and power up. Now you will see that as you move the pendulum left and right, the trolley follows as before—but with a difference. It moves rather more than you move the pendulum, so that the pendulum leans inward—or at least it should if you have the coefficients right. Adjust the coefficient of x so that the pendulum rotates about a point roughly twice as high as the length of the pendulum.

Now give the pendulum another solo run, starting near the center. The trolley will "swing" to and fro with increasing amplitude. Catch it before it hits the ends.

Now add yet another term, some constant times v. When you have the right coefficient, the "swing" will be damped and the pendulum will "rest" near the centre. "Rest" might not really be the right word, since the trolley will jitter to and fro.

You can instead use a multiple of vest. It will actually give a smoother response. So now the only sensors used are the potentiometer measuring x and the sensor measuring tilt. The other variables are deduced from these.

We can go a little further by replacing x in the u= line with (x − xdemand). As the keys are tapped, the trolley will obediently wobble to the left or right as commanded.

You have followed a systematic but empirical process to arrive at feedback coefficients that will stabilize the pendulum. It might have been surprising at first that positive, rather than negative feedback had to be used for the trolley position. In a later chapter, the equations of the systems will be analyzed and your results will be explained.

Another exercise will involve simulating the system. At present we know very few of its parameters, so you should take the opportunity now to measure some.

3.5.5 Measuring the System Parameters

To make an accurate mathematical model of the system, we need to measure a number of parameters. By modifying some of the earlier programs, we can let the computer do most of the hard work for us.

The key parameter is the acceleration of the trolley under full drive. Another is the effect of the velocity on the acceleration.

Which units should we use to measure the position with? The most convenient measure is in terms of the potentiometer voltage. Since we have scaled the ADC output to the range –1 to 1, we will take this range to be two units of position.

Next we need to have some absolute scale of time. We must use the interrupt method in preference to one that relies on the speed of a program loop. The following program will calibrate the interrupt process to give the correct value to use for dt:

```
CLS
PRINT "Calibrating dt - please wait five seconds"

PLAY "mbl64t255"
ON PLAY(1) GOSUB rates
PLAY ON
PLAY "cde"
t = TIMER
n=0
DO
LOOP UNTIL n>=500
PLAY OFF
PRINT "Use dt =";(TIMER - t)/n
END
rates:
n = n + 1
PLAY "n0"
RETURN
```

It simply uses the TIMER function (press ⟨F1⟩ to check it out) to measure the duration of 500 interrupts, which we know to be in the region of five seconds.

Now we need a program to measure acceleration. To test the acceleration, we first ask for the trolley to be moved by hand to the left hand end of the track. When we press ⟨space⟩ for "go", the trolley accelerates under full drive. When it reaches the center, the drive is removed and friction brings the trolley to a halt.

Load in the skeleton you used in Section 3.5.1 and delete even more, then enter the new code. Edit the program that follows to use the dt value that you have just found:

```
'CONSTants required for the ADC routine go here

CONST port= &H378          '(might be 278)
CONST tmax= 1 'Note the change of value
SCREEN 12
WINDOW (0,-1) - (tmax,1)
PLAY "mbl64t255"
dt = .01     'Change this to the value you have found
```

```
ON PLAY(1) GOSUB rates
PLAY ON
PLAY "cde"
'------New code from here on
OUT port, 0
PRINT "Move the carriage to the left end and press
<space>"
DO
LOOP UNTIL INKEY$ = " "          'Wait for key
x0 = ADC(1)
t=0
OUT port, 1                      'Off we go
DO
   PSET (t, x), 14
LOOP UNTIL x>=0 'When we get to the middle
t1 = t          'make a note of the time
x1 = x          'and check the position
OUT port, 0  'switch off and coast to a halt
t=0
DO
   PSET (t, x), 14
LOOP UNTIL t >.9 'Should stop before 0.9 seconds
x2 = x          'How far did it coast?
a = 2 * (x1 - x0) / (t1^2) 'using s=1/2 a t^2
b = a * (x1 - x0) / (x2 - x1) 'v^2 = 2 a s1 = 2 b s2
PRINT "Acceleration = "; a
PRINT "Deceleration = "; b
'----End of new code
DO           'Leave the information on the screen
   a$ = INKEY$
LOOP UNTIL a$ = "q"
PLAY OFF
OUT port, 0
END

rates:
x = ADC(1)
t = t + dt
'Point A
PLAY "n0"
RETURN

FUNCTION adc(chan%)
' The code for the ADC routine goes here
END FUNCTION
```

So, that gives us some numbers. What do they mean?

The fact that the stopping distance is of the same order as the run up suggests that the major slowing effect is not viscous drag but friction. On this assumption we can model the trolley with two equations:

$$dx/dt = v$$

$$dv/dt = cu - \text{friction}$$

where u is +1, −1 or zero according to the drive setting. Friction will be of constant magnitude, multiplied by the sign (±1) of the velocity. Its magnitude will be given by the deceleration that we have just found.

Since the drive has to act on top of the friction, the constant c will be the sum of the acceleration and the deceleration. We are now in a position to make a mathematical model for the trolley movement.

At the point marked `A` in the `rates` subroutine, add

```
vm = vm + (c * u - friction *SGN(vm)) * dt
xm = xm + vm * dt
```

substituting the values you have found for `c` and `friction`.

Now we have to introduce `u` and set the model initial conditions.

At the command `OUT port, 1`, insert

```
u = 1
xm = x
```

and at the `OUT port, 0` command, insert

```
u = 0
xm = x
```

Finally, we must display the model position by adding

```
PSET (t, xm), 12
```

next to the existing `PSET` command. As a result, the new program will show the actual and modeled positions superimposed.

For a more complete model, we need to know more about the pendulum. We can measure the vital parameter with minimal technology. Hang the pendulum upside-down. Give it a swing and measure the period of oscillation.

3.5.6 Final Touches

The program can be polished up for demonstration purposes. The datum value can be written to a file on disk, so that the pole does not have to be balanced at the start.

The original version of the experiment has brackets at the end, forming stops near the midheight of the pendulum, while there are limit stops on the angle to which it can topple. The program can start by applying velocity control to run the trolley gently toward the end. As the pendulum is pushed past the upright position by the stop, the mode switches to balance control and the pendulum is balanced. You can see this in action in a video file to be found at http://www.essmech.com/3/5/6.htm.

A simple test on the tilt angle will detect whether the pendulum has toppled and enable the routine to erect it and start again.

On the new version that is just being commissioned (in 2005), the pendulum can swing freely in a complete circle. The trick now is to erect it from a hanging position by building up oscillations until it can be "caught" at the top.

4

Introduction to the Next Level

A few basic principles and some experimentation can get you a long way in mechatronics, but the time comes when they must be backed up by mathematical theory. That is not to deny that the empirical approach will get you much further than the theory on its own.

I have tried to feature theory that will actually be useful, rather than mental party tricks that are only really relevant to answering examination questions. We cannot avoid differential and difference equations. They are the essential substance of the things that we are trying to control. We must be able to spy out the state variables that define the behavior of the system and then derive state equations to describe them.

One vital use for the differential equations is to enable us to simulate the systems in software and try out the algorithms, whether based on theory or pragmatism, to obtain an early verdict on the likelihood of the success of the outcome. But unless the algorithm is tried out on a real system, simulation is a mere mental pastime.

We have to include methods of testing for stability, for despite the legion of thermostats happy in their limit cycles, there are many systems where true stability is essential. Transfer functions certainly have their uses, but they are just one piece of the jigsaw puzzle and not an end in themselves. When we use mathematical shorthand to put our state equations into matrix form, we find ourselves led into the world of transformations and eigenvalues and we need to brush up on a mathematical toolkit.

Essentials of Mechatronics, by John Billingsley

When we step back to the mechanical reality of moving parts, accelerations and forces, we again find that the coordinate geometry is leading us down the matrix path. Rotations become tensors, and the articulation of a robot arm involves a chain of affine transformations, performed very neatly by a few matrix software operations.

The mechatronic engineer should not be in a hurry to dismiss electronic design too lightly, thinking that purchasing ready-made circuitry is an easy answer. The ADC chip that costs under $10 will find its way into a board marketed for many hundreds of dollars, probably with elaborations that make it difficult to do the simple operations you require. The vendor of the board will offer you a FIFO buffer that can hold hundreds of samples. These are useless to you for control; you need just one value of the variable, measured as recently as possible. You may have to put your own board together!

The ability to throw a circuit together around an operational amplifier will be another essential skill. Once again, a chip costing a few cents will usually be packaged in a smart box with a huge price tag, and there is no guarantee that you can actually find the product that does what you need.

Your circuitry skills will be tested even further if you intend to embed your own microcomputer in your design, rather than exploiting a PC. Think long and hard before you select a particular processor. The wide variation does not just cover price, you must also consider the sizes of program and working memory, number and type of interfaces, and the availability of cross-support software. Many devices such as the PIC have a huge hobbyist following, with an abundance of freeware available on the Web. Others can include bus systems aimed at industrial applications such as the motor industry, but these too may have excellent free cross support.

4.1 THE www.EssMech.com WEBSITE

There are things that a book simply cannot achieve, such as real-time simulations, interactive examples, and movie illustrations. Those shortcomings are easily overcome by mounting a Website for readers of the book.

4.1.1 Examples and Simulations

There is nothing like trying out a simulation in real time to see the problems that a control system can really present. But what is the easiest way to present the simulation experience to the reader? Packages seldom come cheaply. They also present problems all of their own in the "learning curve" requiring time to use them to their full advantage.

There is a graphics computing environment already installed on your computer. It is hidden within the browser. It allows you to enter or modify code displayed in a text area of a Web page and execute it at the click of a button.

The language is JavaScript, and it is used here with a simple Java "applet" that puts all the graphic tools within reach. With the title *Javascript On-Line Learning Interactive Environment for Simulation*, examples have already been on show for some time at http://www.jollies.com/.

Although the code looks very much like C, there are some subtle differences. Nevertheless, it is easy to edit or create new code for the Jollies pages both for simulations and for graphic image processing.

Although the Jollies are probably the easiest route by which to approach the examples, many on the EssMech Website are duplicated in QBasic or QuickBasic (the syntax is the same) and in Visual Basic as well.

You will find other code there, too. There are examples of assembly code for embedded microcomputers and "solution" code for the laboratory exercises. That is in addition to some packages to help with vision examples and experiments and the occasional movie of mechatronics in action.

4.1.2 Finding the Code

If you simply enter the URL of the Website, you will find pages that tell you about the book, with links to the examples.

To find a direct path to each example, you must add the subsection number in a slightly cryptic way as follows; The example for this subsection, 4.1.2, would be linked at http://www.essmech.com/4/1/2.htm. Try it! You can also try http://www.essmech.com/4/1/ to see an index of examples in the whole of Section 4.1.

5

Electronic Design

Knowing nothing more than the rudiments of circuit theory, it is possible to use catalog components to design amplifiers, filters, discriminators, and even an elementary analog-to-digital converter.

Those rudiments must include knowledge of how to calculate values for the combination of passive components (resistors, capacitors, inductors) in series and parallel and the analysis of circuit loops by Kirchhoff's methods. Familiarity with Norton and Thevenin's theories would also help.

The catalogs are full of semiconductor devices, many costing no more than 10¢ and few costing more than $5. They range from elementary amplifiers and logic circuits, through comparators, counters, multichannel ADCs, power transistors, magnetic and optical sensors, and many of the embeddable microcomputers.

In the same catalogs you are likely to find plug-in systems to solve your problems. These, however, are likely to cost a hundred times as much. Being designed to address the problems of a hopefully large client base, it is unlikely that they will be the best fit for your specific application.

It is well worth gaining expertise in putting your own circuits together.

5.1 THE RUDIMENTS OF CIRCUIT THEORY

Circuits can be arranged as networks, as in Fig. 5.1, nodes joined by meshes to form loops. The laws and equations can be expressed in many ways, but can be summed up as described in the following paragraphs.

Essentials of Mechatronics, by John Billingsley

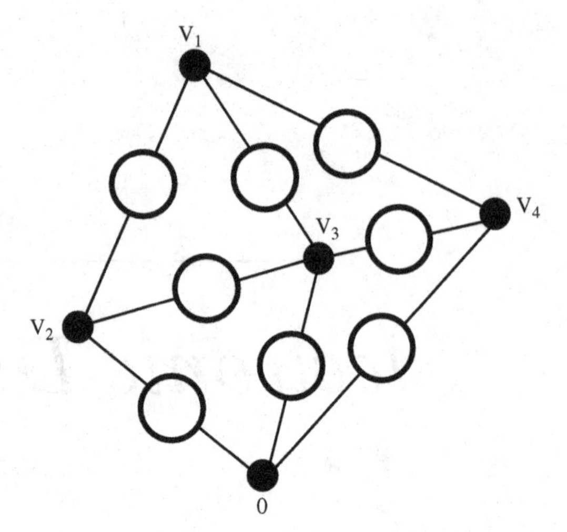

Figure 5.1 *Circuit network.*

Each node has a voltage with reference to the node 0 that is taken as reference.

Each mesh carries a current between nodes, taking the direction and sign into account. The first and only rule states that the sum of all currents at each node must be zero. In other words, all currents must "come from somewhere." (Kirchhoff's "second rule" is a simple consequence of assigning a voltage to each node.)

To analyze the circuit, we must therefore have some way of relating the current in a mesh to the difference between the voltages of the nodes at its ends. We have a number of primitive components for this analysis:

1. *A Voltage Source.* This will be a voltage that contributes to the voltage difference across the mesh regardless of the current passing through it.
2. *A Current Source.* Rather more aggressive, perhaps, the current source forces its current in the mesh regardless of the voltage across it. Clearly the resulting equation will express the voltage across the mesh in terms of this current, rather than vice versa.
3. *Resistance.* Ohm's law states that the voltage across the resistance is proportional to the current through it, but this is more a rule of thumb than a law of physics. With a wide variation in current resulting in temperature changes in the components, nonlinearities are likely to become apparent. As soon as semiconductors are involved, the nonlinearities become important.

   Resistance "contaminates" most other components, so that they come in combination with it. A voltage source is seldom "pure," but appears to have a resistance in series with it. A current source will usually appear to have a resistance across it.

4. *Capacitance.* Capacitance has a somewhat different nature. It involves time. The current through a capacitor is proportional to the rate of change of the voltage across it, so the equation for a mesh is then likely to turn out to be a state equation.
5. *Inductance.* Inductance also involves time, but this time the voltage across the inductance is proportional to the rate of change of the current through it. Once again, an inductance is seldom "pure" but usually has some resistance associated with it.

Electricians have assembled a set of rules for combining components in series and parallel. When current passes through two resistors in series, the voltage across them is, of course, the sum of the two individual voltages. The combined resistance is thus the sum of the two resistances:

$$R = R_1 + R_2$$

When they are connected in parallel, both resistors have the same voltage across them and pass the sum of the two individual currents:

$$\frac{1}{R} = \frac{1}{R_1} + \frac{1}{R_2}$$

Using these two simple rules, a circuit consisting only of resistors can be "crunched" first by working out a single resistance for each mesh, then combining meshes in parallel.

Thevenin's theorem states that a circuit, however complicated, can be reduced to a single resistance in series with a single voltage source. Norton turns this on its side and says that this is equivalent to a single resistance (of the same value) in parallel with a single current source. Both theories, of course, depend on the circuit components being linear.

Flushed with their success at manipulating resistors, electricians are keen to represent capacitors and inductors in the same form. Often their calculations will involve a single supply frequency, which they multiply by 2π to get the *angular frequency*, ω. The sines and cosines of the waveform can be expressed as the real and imaginary parts of $e^{j\omega t}$, and every differentiation will give rise to a multiplying factor of $j\omega$. They call an inductance L an *impedance* of $Lj\omega$ and crunch it just as if it were a resistance.

The ratio of output to input voltages of a network is likely to involve a spattering of $(j\omega)$ symbol, and the result is called a *transfer function*. Their euphoria with transfer functions is carried over into their dealings with control systems, and their influence is to blame for the tradition of treating transfer functions as the foundation for the teaching of control theory.

One particular form of network of particular interest is the *two-port* or *four-terminal* network (Fig. 5.2). The two output terminals can be tied to the

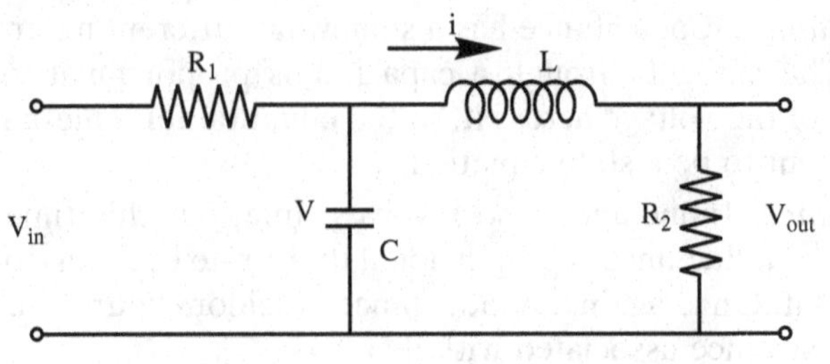

Figure 5.2 Four-terminal network.

two input terminals of another such network; indeed, a whole chain of networks can be linked to form a complex filter.

Four variables are involved, the input voltage and current and the output voltage and current. The equations linking them can be found by "network crunching," or more simply by building state equations. Consider the circuit of Figure 5.2.

The state variables are V, the voltage on the capacitor; and i, the current through the inductor. The voltage across R_1 is $V_{in} - V$, so the current through it is

$$\frac{V_{in} - V}{R_1}$$

The current through the inductor is, of course, just i, but is directed away from the capacitor. The currents arriving at the top node of the capacitor, including the current passing through the capacitor, must sum to zero. The current flowing into the capacitor is equal to C times its rate of change of voltage:

$$C\frac{dV}{dt} = \frac{V_{in} - V}{R_1} - i$$

Now the current through R_2 will be $i - i_{out}$, and the voltage across it is

$$R_2(i - i_{out})$$

so we can find the voltage across the inductor. The equation that defines an inductance gives us

$$L\frac{di}{dt} = V - R_2(i - i_{out})$$

and there we have our two state equations. Everything on the righthand side is either a state variable or an input—note that both V_{in} and i_{out} are inputs, as far as this system is concerned.

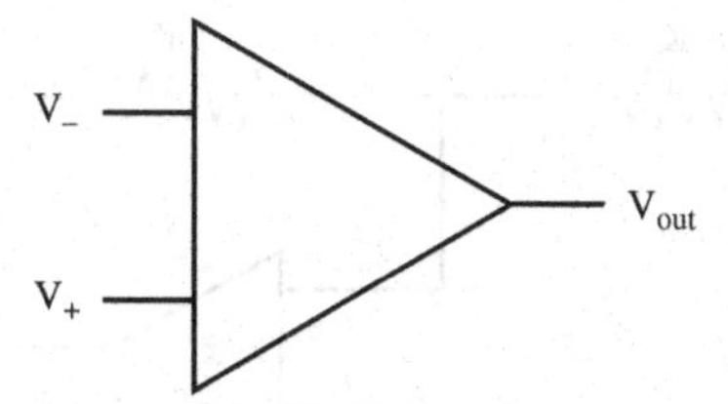

Figure 5.3 *Operational amplifier.*

5.2 THE OPERATIONAL AMPLIFIER

This type of integrated circuit seems as versatile as the transistor itself, but allows amplifier and signal conditioning circuits to be designed to a "grand scheme" with few of the niggling details of biasing that transistors present.

The output voltage V_{out} (see Fig. 5.3) is a large number of times the difference between the two input voltages—where the large number may typically be 100,000. But it is most unlikely that you will want such a high gain in your final circuit. The gain is reduced when you apply feedback around the amplifier, in exchange for certainty about the closed-loop circuit parameters.

Modern op-amps (operational amplifiers) have a very high input impedance. That means that when you apply a voltage to the input, practically no current at all is accepted. They also have a very small "input offset voltage," the actual difference in voltage between the inputs when the output is zero.

5.2.1 Virtual Earth

A large family of applications use a "virtual earth" concept, where the non-inverting input of the amplifier is connected to ground, or to a constant reference voltage.

Now let us start to analyze the circuit, even before we decide what the circuit will actually consist of. Since

$$V_{out} = A(V_+ - V_-)$$

and since we have grounded V_+, we know that

$$V_- = -V_{out}/A$$

and since A is very large, that will make V_- virtually zero, unless the amplifier is saturated. Hence the term "virtual earth."

Now, for example, if we connect a resistor of 100 kΩ between V_{out} and V_- (as shown in Fig. 5.4), both V_{out} and V_- will be zero. Let us connect a second resistor to V_-, this time of value 10 kΩ, and apply one volt to the end of it. What does this do to V_-?

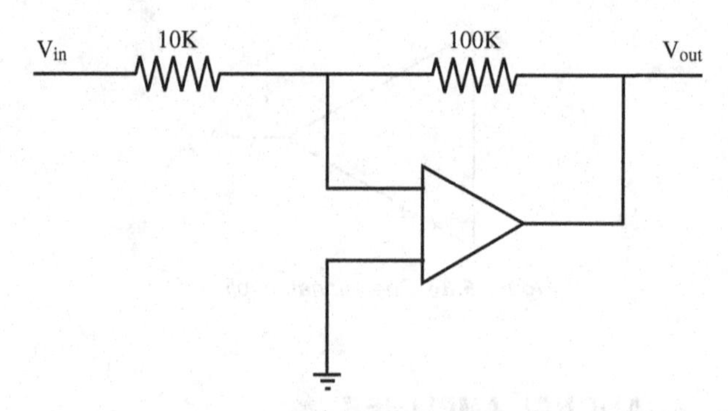

Figure 5.4 *Amplifier with virtual earth.*

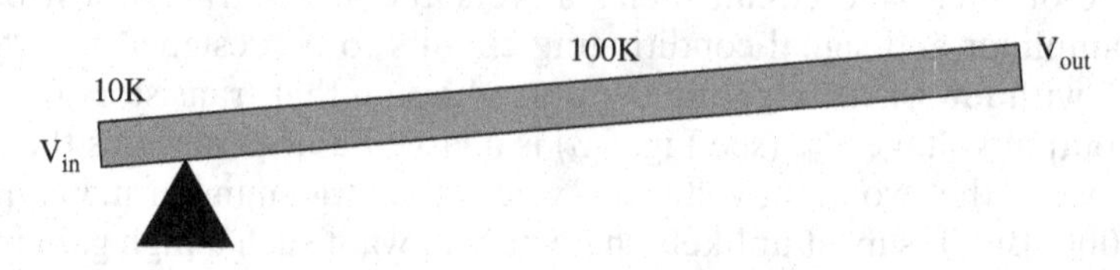

Figure 5.5 *Amplifier seesaw analogy.*

As soon as V_- is pulled away from zero volts, V_{out} changes to try to pull it back. By taking a value of –10 V, the two resistor currents cancel out and balance is restored. We have made an amplifier with a gain of –10; thus, for every volt applied at the input to the 10 kΩ resistor, the output will change by –10 V.

You can look on the schematic rather as a seesaw (Fig. 5.5), with one side of the plank 10 times as long as the other.

Let us now connect an additional 10 kΩ resistor to V_- and apply –2 V to it. Now the sum of all three currents at the V_- junction must be zero, while the voltage there remains at zero. The output must change to +10 V. We have devised a way to add input signals together, although those signals do not "see each other" since they are only joined via a point that remains at zero volts. The virtual-earth connection has become a "summing junction" (see Fig. 5.6).

We can add signals in unequal proportions by varying the values of their corresponding resistors. If we want to subtract a signal, we can first invert it using an operational amplifier with an input resistor equal to its feedback resistor.

If we replace the feedback resistor with a capacitor, we get an *integrator* (Fig. 5.7). As before, the currents at the summing junction must add to zero. Now the equation that describes the current i in a capacitor is

$$i = C\frac{dV}{dt}$$

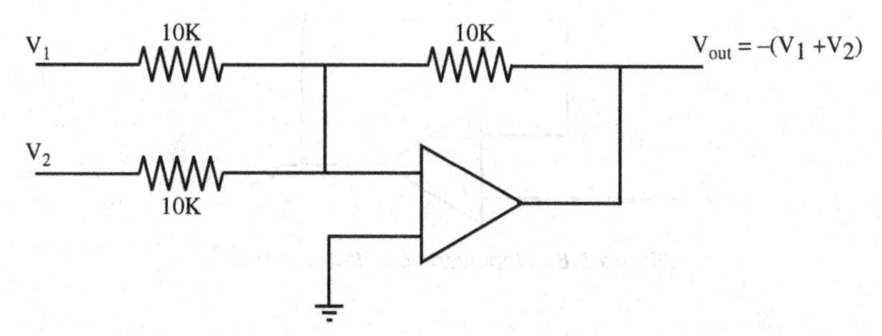

Figure 5.6 *Summing amplifier.*

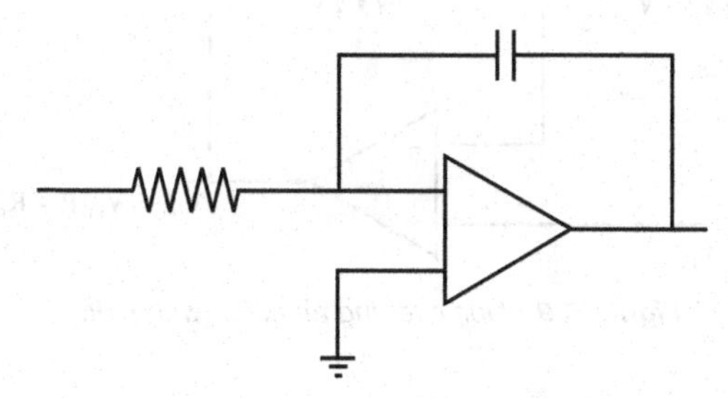

Figure 5.7 *Integrator with capacitor feedback.*

So we have

$$\frac{V_{\text{in}}}{R} + C\frac{dV_{\text{out}}}{dt} = 0$$

or in other words

$$\frac{dV_{\text{out}}}{dt} = -\frac{V_{\text{in}}}{RC}$$

The derivative of V_{out} is proportional to V_{in}, so V_{out} is proportional to the integral of V_{in}.

Now we have all that we need to make an analog computer. With it, we can simulate linear systems, or add some circuit dodges to simulate nonlinear ones. This was the method of choice in the mid 1970s, but today it is so much easier to simulate a system digitally.

This does mean, however, that with some operational amplifiers and a handful of capacitors, we can produce virtually any transfer function. This could be all that we need to stabilize a difficult system.

5.2.2 Other Configurations

Instead of tying V_+ to ground, we can apply the input signal to it (see Fig. 5.8). The virtual-earth configuration has the disadvantage that the amplifier

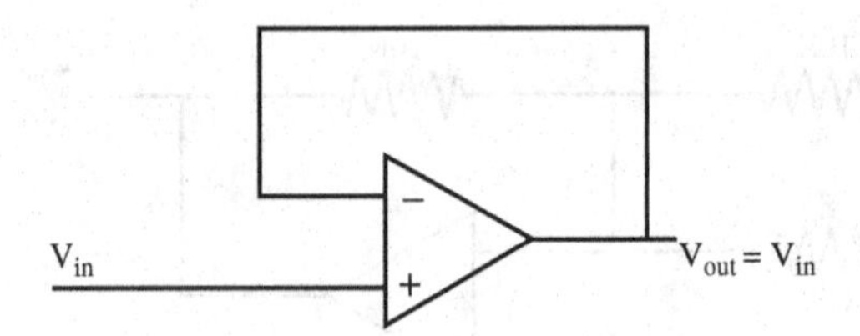

Figure 5.8 Noninverting buffer amplifier.

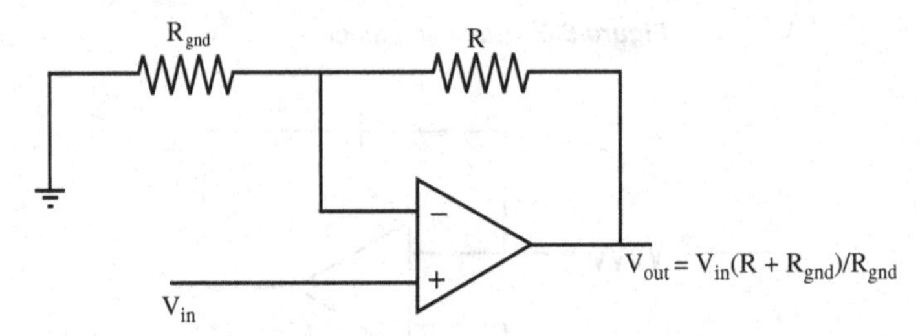

Figure 5.9 Noninverting amplifier with gain.

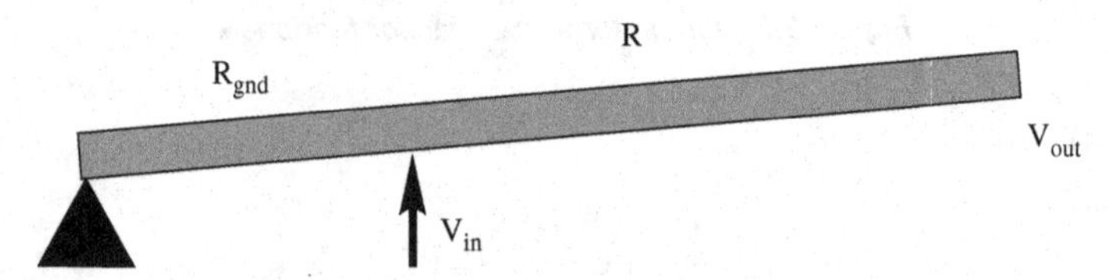

Figure 5.10 Seesaw analogy becomes a lever.

has an input impedance that is no greater than the input resistor. Some signal sources should not be loaded, even by this sort of resistance value.

With the signal applied to the noninverting input, however, the input impedance is extremely high. Via the feedback, the output will drive the inverting input to match V_+, so the input impedance will be many times larger than that of the op-amp itself.

To avoid any risk of oscillation, it may be preferable to apply this feedback via a resistor, rather than a direct connection. By connecting a second resistor between V_- and ground, we can have high input impedance combined with some noninverting gain (Fig. 5.9). The "seesaw" principle becomes a lever (Fig. 5.10) and the gain is seen to be $(R + R_{gnd})/R_{gnd}$.

The circuit shown in Figure 5.9 is the same one we used in the ball-and-beam experiment. When we add a capacitor between V_+ and ground, if the ball has broken contact with the track, the voltage on V_+ will decay very slowly due to the high input impedance. As soon as the ball makes contact with the track again, the effect of the capacitor is only a very short time constant.

As with the integrator, the time constant is given by the value of *RC*. If our capacitor is chosen to be one microfarad and if the track resistance is 10 Ω, the result will be 10 μs, which is much too small to have any effect on the system's performance.

5.2.3 Differential Amplifier

Some signals, such as strain-gauge outputs, appear as small differences between two voltages that are not close to ground. Other signals might have mains noise superimposed on the pair of leads that we are trying to measure. In these and many other cases we would like an amplifier that has good *common-mode rejection*, which will amplify the difference voltage and not respond to signals that vary both lines together.

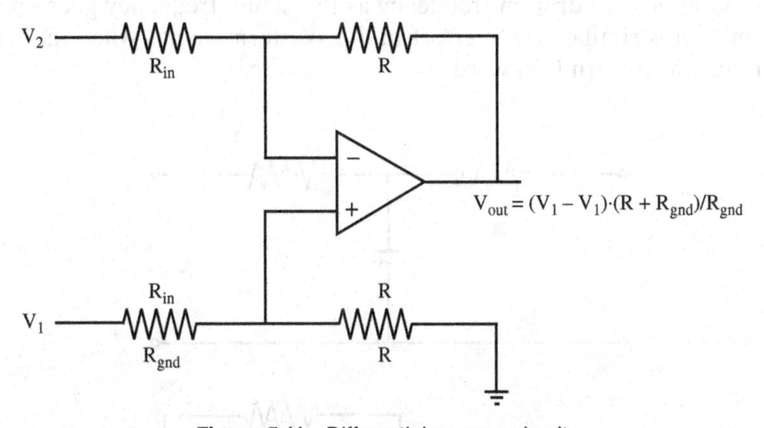

Figure 5.11 Differential op-amp circuit.

The differential circuit shown is Figure 5.11 is the answer. If we ground input *A* and vary input *B*, we see that the "seesaw" gain of R_2/R_1 applies to the difference of the inputs. But if we tie inputs *A* and *B* together and vary them, we see that the voltages on V_+ and V_- remain equal if V_{out} remains at zero. The common-mode gain is thus zero.

5.3 FILTERS FOR SENSORS

In addition to amplifying sensor signals, we may need to process them in other ways. One requirement might be to remove high-frequency noise.

5.3.1 Antialiasing

It is, of course, possible to take averages of digitized readings to smooth out some types of noise, but the digitizing process itself suffers from *aliasing* (Fig.

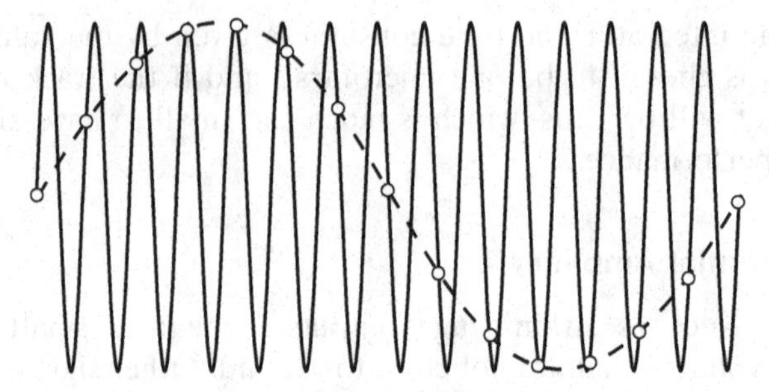

Figure 5.12 *Example of aliasing.*

5.12). As frequencies rise above half the sampling frequency, the sampled signal can appear to drop in frequency as the actual frequency goes up. The phenomenon is similar to the effect in old Western movies when the wagon wheels appear to turn backward.

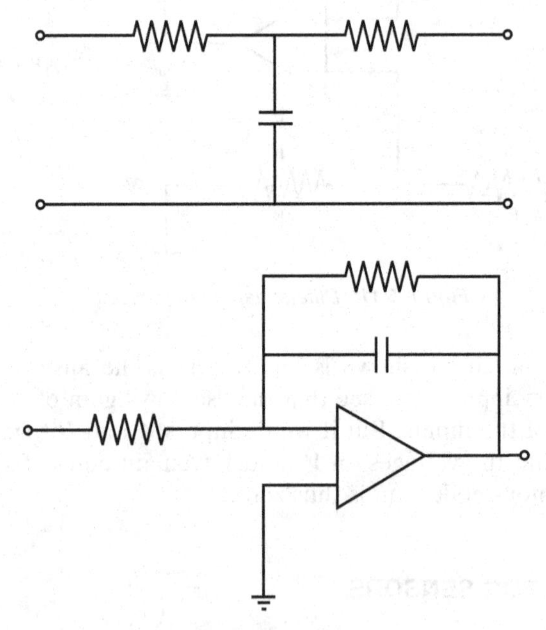

Figure 5.13 *Two simple lowpass filters.*

The only way to eliminate the high-frequency noise is to attack it before digitizing, using an *antialiasing filter*. This can simply be a lowpass *RC* circuit (e.g., see Fig. 5.13), an op-amp with feedback consisting of a resistor in parallel with a capacitor, or a higher-order filter with several capacitors.

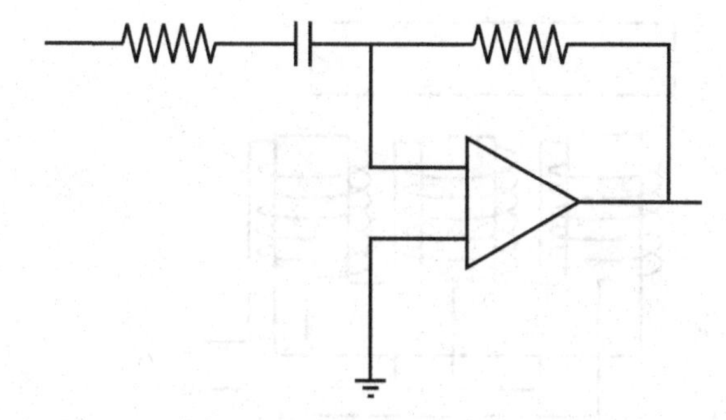

Figure 5.14 Differentiator circuit.

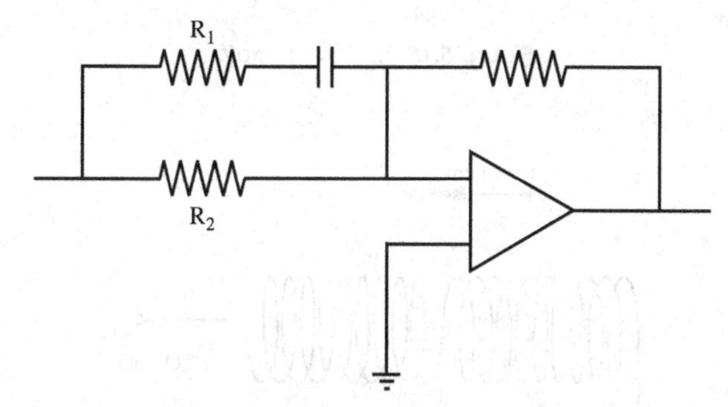

Figure 5.15 Phase advance circuit.

5.3.2 Differentiating and Phase Advance

In principle we could swap the resistor and capacitor of an integrator to form a differentiator. In practice, the gain increases indefinitely at high frequencies, so the output would be swamped with noise. We have to add a resistor in series with the capacitor (see circuit in Fig. 5.14). The output is a lowpass version of the derivative, with a limit on the high-frequency gain.

When we add this estimated derivative to the original signal, as when adding an estimated velocity to a position signal, the result is a *phase advance* (see circuit in Fig. 5.15).

At high frequencies, the capacitor can be regarded as a short circuit, while at DC it acts as an open circuit. The ratio between the gains at high and at low frequencies is thus $(R_1 + R_2)/R_1$.

5.3.3 Switched Filters

Possibly the signal that we seek lies is modulated by an alternating voltage. Many sensors use AC signals, such as the *E and I pickoff* (Fig. 5.16) used in the past for aircraft instrumentation and the *linear variable differential transformer* (LVDT) (Fig. 5.17). The output is an AC signal that is zero at a central

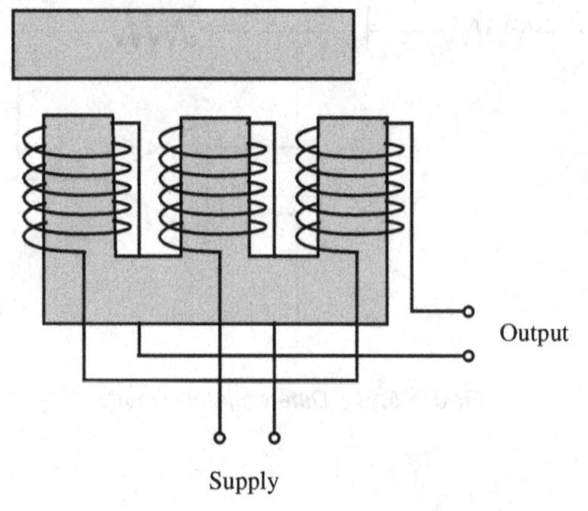

Figure 5.16 E and I pickoff.

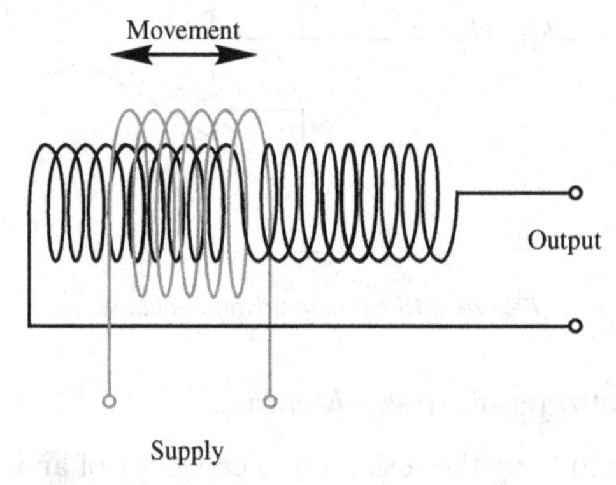

Figure 5.17 Linear variable differential transformer.

position, increasing with opposing phases as the sensor is displaced on either side of the center.

To convert such a signal to a DC one that can vary positive and negative, we require a *phase-sensitive discriminator*. For this, we can use a semiconductor switch.

A circuit such as the CMOS 4066 depends on field effect transistors to close or open connections between pairs of contacts. The control signal is designed to be operated by the output of a PC logic line or a microprocessor, but here we use the oscillator supply for the switching signal (see Fig. 5.18).

5.3.4 A Single-Chip ADC

In Chapter 3, many of the experiments depended on the availability of an ADC. If you have no suitable card, a four-channel or eight-channel ADC can

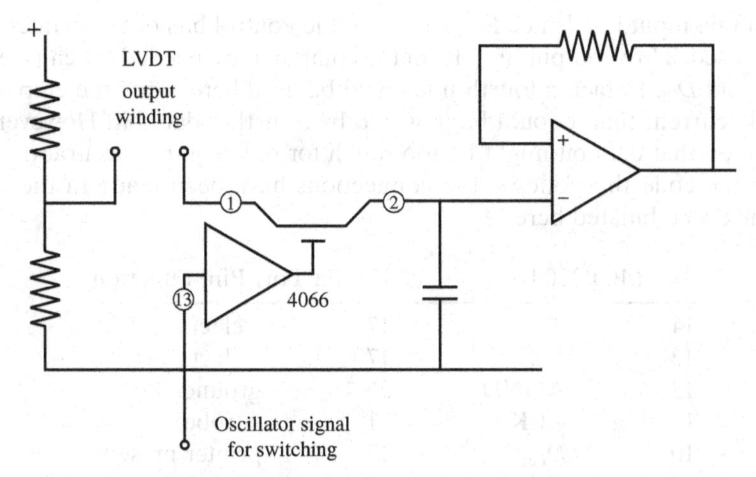

Figure 5.18 Half-wave switched discriminator.

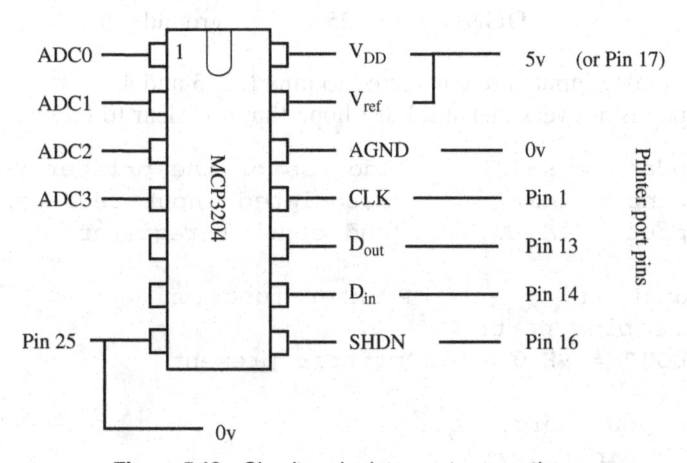

Figure 5.19 Circuit and printer port connections.

be built from a single chip, connected to the parallel port of the PC with no additional electronic components.

The MCP3204 and MCP3208 chips from Microchip Technology Inc. are designed to communicate using a serial technique. Data bits are clocked in and out in response to a clock signal generated by the computer to which the chip is connected.

When the "shutdown" line is pulled to ground, the chip is selected. Data on the D_{in} line is clocked into the chip to set the channel number to be converted. After an extra clock cycle, the chip starts to output the 12-bit result on the D_{out} line, most-significant bit first. When the 12th bit has been received, the computer sets the shutdown line high to reset the chip before the next conversion. (See configuration in Fig. 5.19.)

Just one input line is needed, so one of the control bits of the printer port can be used. Three output lines from the computer are needed for chip select, clock, and D_{in}. In fact, a fourth line could be used here, since the chip takes so little current that it could be powered by a further data bit. However the voltage on that data bit might be too rough for our required accuracy.

For the code that follows, the connections have been made in the same sequence as tabulated here:

MCP3204		Printer Port Pin, Function	
14	V_{DD}	17	-select
13	V_{ref}	17	-select
12	AGND	25	ground
11	CLK	1	-strobe
10	D_{out}	13	printer present
9	D_{in}	14	-auto LF
8	SHDN	16	initialize
7	DGND	25	ground

The four analog inputs are connected to pins 1, 2, 3 and 4.

The code is not very elegant, but I hope that it is clear to follow.

```
CONST pdata = &H378      'Address of the printer port
CONST pinp = &H379       'associated input register
CONST pout = &H37A       'and control register

'MCP function            Printer function
' On the pinp port:
CONST DOUT = &H10        'printer present

'On the pout port:
CONST CLKbar = 1         '-Strobe
CONST DINbar = 2         '-auto linefeed
CONST SHDN   = 4         'initialize
CONST VDDbar = 8         '-select

CONST CLKlo = CLKbar + DINbar 'VDD high, shutdown low
CONST CLKhi = DINbar          'same with CLK high

'Demonstration program:
CLS
SCREEN 9
WINDOW (0, -1)-(1000, 1)
DO
   FOR i = 0 TO 1000
      PSET (i, ADC(0))
```

```
   NEXT
CLS
LOOP UNTIL INKEY$ <> ""

FUNCTION ADC (chan%)
DIM i AS INTEGER, bits AS INTEGER

bits = 0

OUT pout, CLKlo             'VDDbar=0, shutdown=0

OUT pout, CLKlo - DINbar  'Start bit must be 1
OUT pout, CLKhi - DINbar  'clock it
OUT pout, CLKlo - DINbar  '(subtract DINbar to set DIN
                           high)

OUT pout, CLKlo - DINbar  'First bit 1 for single ended
OUT pout, CLKhi - DINbar  'clock it
OUT pout, CLKlo - DINbar

OUT pout, CLKlo  'For 3204, don't care - so output zero
OUT pout, CLKhi  '(for 3208, same as MSB of address)
OUT pout, CLKlo
IF (chan% AND 2) THEN      'High bit of channel number
   OUT pout, CLKlo - DINbar
   OUT pout, CLKhi - DINbar
   OUT pout, CLKlo - DINbar
ELSE
    OUT pout, CLKlo
    OUT pout, CLKhi
    OUT pout, CLKlo
END IF

IF (chan% AND 1) THEN      'Low bit of channel number
    OUT pout, CLKlo - DINbar
    OUT pout, CLKhi - DINbar
    OUT pout, CLKlo - DINbar
ELSE
    OUT pout, CLKlo
    OUT pout, CLKhi
    OUT pout, CLKlo
END IF

    OUT pout, CLKlo
    OUT pout, CLKhi   'The chip samples input now
```

```
    OUT pout, CLKlo
    OUT pout, CLKhi     'extra clock was needed

    OUT pout, CLKlo     'null bit starts now
    OUT pout, CLKhi

FOR i = 0 TO 11
    OUT pout, CLKlo     'next bit ready now, MSB first
    bits = 2 * bits + (INP(pinp) AND &H10) \ &H10
    OUT pout, CLKhi
NEXT

ADC = (bits XOR 2048) / 2048! - 1
OUT pout, SHDN + CLKbar + DINbar     'shutdown high

END FUNCTION
```

5.3.5 A More Rudimentary ADC

For serious applications, it would probable be advisable to use a commercial ADC chip or to use the ADC capabilities of an embeddable microcomputer. To support the experiments of Chapter 3, however, it is possible to construct a converter from very simple components and use the power of software. That, after all, is part of the essential art of mechatronics.

The heart of the device is a capacitor that charges at a steady rate, to give a ramp of voltage. An output bit from the computer first drives a transistor switch that discharges the capacitor and holds it at zero voltage. An output command frees it to ramp upward to approach the supply voltage. Four comparators on a single chip compare the ramp against three inputs and a reference voltage. These outputs are connected to the bits that serve as input bits on the printer port, where they are monitored in a counting loop. The loop ends when the ramp passes the reference voltage. The circuit is shown in Figure 5.20.

Take care when ordering components. The traditional package for small integrated circuits was *dual inline* (DIL). Here the two rows of pins are spaced at 0.1 in. intervals and the component can easily be mounted and soldered in a *project board*. For hand assembly, you must be sure to obtain DIL components.

Manufacturing has now turned almost completely to *surface-mount* technology, where, instead of pins, the component has a row of small metal pads that bond to the surface of the printed-circuit board when dabs of solder paste are heated and melted.

If you have faith in your circuit skills—and if your computer is not a cherished favorite—you can obtain 5 and 12 volt supplies from inside it. Take them from one of the spare power connectors intended for plugging into disk units.

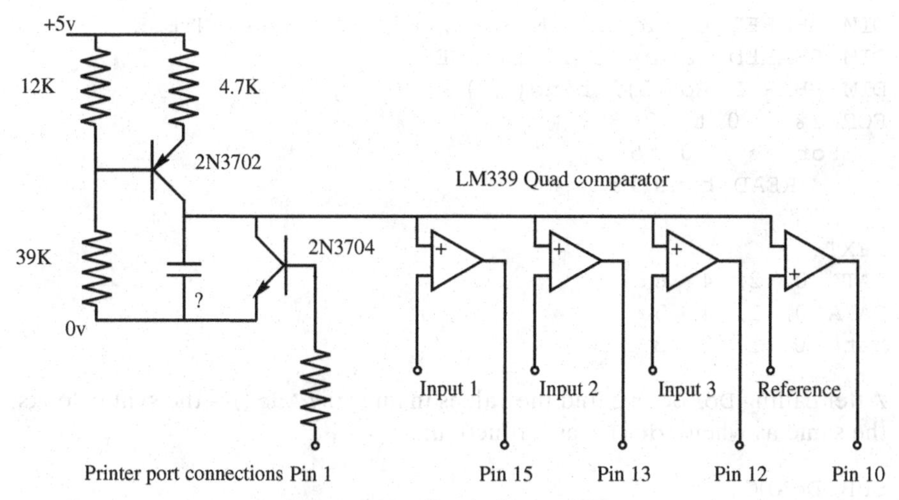

Figure 5.20 *Circuit of ramp ADC.*

As always, there are drawbacks. The PC, even in DOS mode, is interrupted frequently to service "housekeeping" interrupts, including updating the clock that controls the TIMER function. If one of these interrupts happens during the ramp, the wrong answer will be given. However, the count given for the reference voltage will also be less that it should, so a second attempt can be made or the previous value retained.

The software is as follows. First we have the three addresses for the printer port and the bit definitions, corresponding to the connector pins:

```
CONST pdata = &H378
CONST pinp = &H379
CONST pout = &H37A

' On port pinp:
CONST pin15 = 8           'True if high, used for error
CONST pin13 = &H10        'True if high, printer present
CONST pin12 = &H20        'True if high, out of paper
CONST pin10 = &H40        'True if high, ack
CONST pin11 = &H80        'True if low,  -busy

'On port pout:
CONST pin1 = 1            'True gives low , Strobe
CONST pin14 = 2           'True gives low, Auto linefeed
CONST pin16 = 4           'True gives high, initialise
CONST pin17 = 8           'True gives low, select
```

We define some shared variables and set up some other values:

```
DIM SHARED count AS INTEGER, mincount AS INTEGER
DIM SHARED Table(15)AS INTEGER
DIM SHARED Adc(3), bins(3,3)
FOR i% = 0 to 2
   For j% = 0 to 3
      READ bins(i%, j%)
   NEXT
NEXT
DATA 0, 2, 4, 6
DATA 0, 1, 4, 5
Data 0, 1, 2, 3
```

After calling DoADC, we find the values in an array Adc()—the syntax looks the same as when Adc() was a function:

```
SUB DoADC
FOR i% = 0 to 15
   Table(i%) = 0
Next
count = 0
OUT pout, pin1                                'release ramp
DO
   x% = INP(pinp)\8                           'Whole number
divide
   Table(x% AND 15) = count
   count = count + 1
LOOP UNTIL x% AND pin10                       'ramp crosses
                                              reference

OUT pout, 0                                   'reset ramp
IF count < mincount THEN EXIT SUB             'leave old results
FOR i% = 0 TO 2
   k% = Table(0)
   FOR j% = 1 TO 3
      IF Table(bins(i%, j%)) > k% THEN
         k% = Table(bins(i%, j%))
      ENDIF
   NEXT
   Adc(i%) = 2 * k% / count - 1
NEXT
END SUB
```

So, how does it work? Before the ramp crosses the reference voltage, each time we input from pinp and divide by 8, we will get an answer between 0 and 7. We store the value of count in the corresponding bin. When we have finished "filling the bins," we look at what we have caught.

Any time that the ramp has not crossed input channel 0, the first bit of the bin number will not be set. The bin number will therefore be 0, 2, 4, or 6. So, if we look for the largest value in these four bins, we will find the value of count just before the status of pin 15 changed.

Similarly, the value for channel 1 will be given by the largest value in bins 0, 1, 4, and 5, where the bit of value 2 in the bin number is not set.

Now all that remains is to scale the answer to the range –1 to 1 and store the result in the `Adc` array.

Before starting any interrupts, we should call `DoAdc` several times and note the largest value of count that it gives. Reduce this by, say, 5%, and save the answer in `mincount`. Since we will always `PLAY "n0"` between calls, the ramp should have had time to return safely to zero.

You have a further task to complete. You must choose an appropriate value for the ramp capacitor. Since we are using a software loop for timing, this will depend on the computer speed and on whether you are using QBasic, Quick Basic, or a compiled .EXE file. Ideally you want `count` to be about 1000, to give a resolution of 0.1%. Too low, and you lose accuracy; too high, and you risk an overflow error if `count` ever exceeds 32,767. Start small, run the program with a `PRINT count` line, and dab on extra capacitors in turn until you find something close to what you want.

5.4 LOGIC AND LATCHES

There have been several generations of changing technology—including RTL, DTL, TTL, and CMOS—of the extensive family of logic circuits. Now the 7400 and 4000 series CMOS seem to be winning the day, but the variations are legion. Do you need "high speed," "very high speed," "low voltage," "advanced," or "Schottky"? A 2004 catalog contained 15 pages with well over 100 devices listed per page.

It can be shown mathematically that every possible logic circuit can be constructed from a combination of NOR gates. What is a NOR gate? If either of its two logic inputs is at a logic 1, near 5 V, the output will be at logic 0 near ground. Only if neither one input NOR the other is a 1 will the output be high. It can be described by a "truth table" as shown on the right in Figure 5.21.

In the days of *resistor–transistor logic* (RTL), such a gate was constructed from just one transistor and three resistors (see Fig. 5.22).

Two NOR gates can be connected together to make a 1-bit memory (Fig. 5.23). If the two free inputs are low, then either output *A* is high, forcing output *B* to be low, hence allowing *A* to be high—and so on, round in a circle, or equally output *B* can be high forcing *A* to be low, and so on. If input *a* becomes high, *A* will be forced low and *B* will become high. When *a* falls again, the outputs remain in the state they were left. If output *b* becomes high for a moment, the output will "flip" into the other state—the circuit is sometimes known as a "flip-flop".

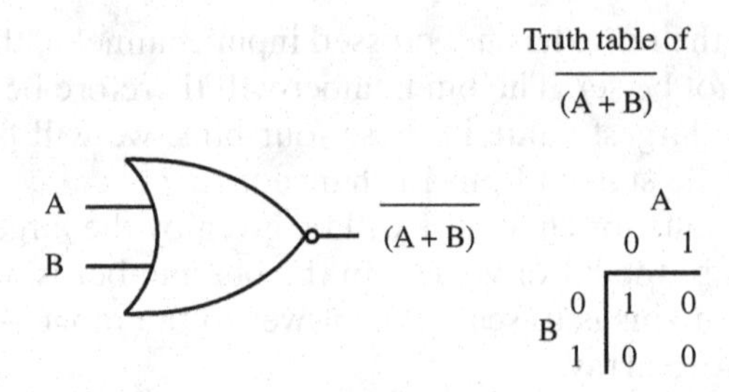

Figure 5.21 NOR gate and truth table.

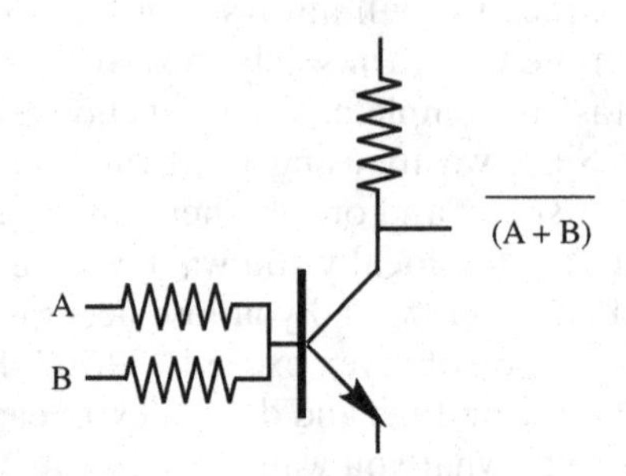

Figure 5.22 Transistor NOR gate.

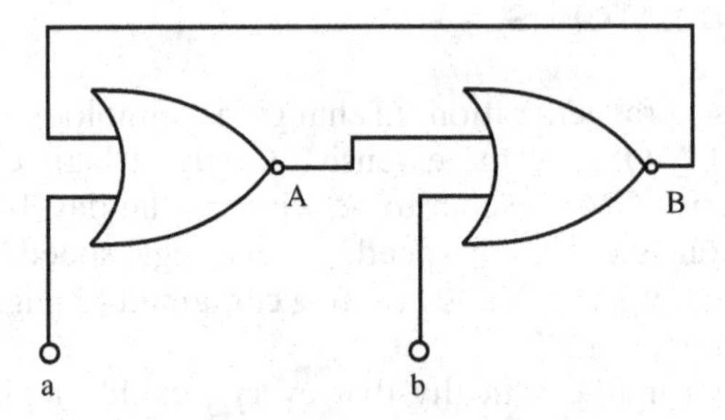

Figure 5.23 Two NOR gates as a flip-flop.

Today, serious circuits are more likely to use one of the mass of "latches" or *JK* flip-flops that fill the catalog, four or more to a chip. Even these are not likely to be found beyond a "breadboard" experimental circuit, since entire systems can be implemented by a *field-programmable gate array* (FPGA) or burned onto an *application-specific integrated circuit* (ASIC).

Logic can be analyzed by *Boolean algebra*, by which expressions such as

$$(A \text{ and not}(B \text{ or } C)) \text{ or } ((\text{not } A) \text{ and } (\text{not } B) \text{ and } (\text{not } C))$$

can be simplified for easier circuit design. Some essential rules can be deduced from common sense:

$$A \text{ and } (B \text{ or } C) = (A \text{ and } B) \text{ or } (A \text{ and } C)$$
$$A \text{ or } (B \text{ and } C) = (A \text{ or } B) \text{ and } (A \text{ or } C)$$
$$\text{not } (A \text{ or } B) = (\text{not } A) \text{ and } (\text{not } B)$$
$$\text{not } (A \text{ and } B) = (\text{not } A) \text{ or } (\text{not } B)$$

while

$$B \text{ or } (\text{not } B) = \text{true}$$

In some computer languages, logic operators are expressed as words such as these, while others use symbols such as &, |, and !. Mathematicians grace the subject with the term *propositional calculus* and use a host of other symbols.

As an exercise, try simplifying the first expression above.

For mechatronics, logic design is likely to be a means to the end of including logic states in the controller. If there is a computer involved, it is usually easier to sidestep the problem of logic design and use software.

6

Essential Control Theory

Control theory is traditionally taught from the point of view of the frequency response, with great emphasis on the manipulation of transfer functions. Instead we will start with the *state space* approach, based on differential equations that you can identify from the "real" system.

6.1 STATE VARIABLES

The relationship between state variables and *initial conditions* has already been mentioned, but let us try to make the concept as clear as possible.

A cup of coffee has just been prepared. It is rather too hot at the moment, at 80°C. If left for some hours, it would cool down to room temperature at 20°C, but just how fast would it cool, and when would it be at 60°C?

The cup remains full for now, so just one variable interests us: T, the temperature of the coffee. It is a reasonable assumption that the rate of fall in temperature is proportional to the temperature above ambient. So we see that

$$\frac{dT}{dt} = -k(T - T_{\text{ambient}})$$

If we can determine the value of the constant k, perhaps by a simple experiment, then the equation can be solved for any particular initial temperature—although we'll look at the form of the solution later.

The concept of state variables is so simple, yet it is essential for gaining insight into dynamic systems. As an exercise, consider the following systems, select state variables, and derive state equations for them:

1. The money in a bank account that carries compound interest.
2. The voltage on a capacitor that has a resistor connected across it—assuming that it is originally charged.
3. The distance of the back wheel of a bicycle from a straight line when the front wheel is wheeled along the line—assume that it starts away from the line.
4. The speed of a motor when driven from voltage V.

Think up your own answers before reading on:

1. The state variable in this case is just your credit balance. To find its rate of change, multiply the credit balance at this very instant by the interest rate. If we call the credit c and the interest rate R, then the equation is just

$$\frac{dc}{dt} = Rc$$

where, of course, time is measured in years.
2. This time the state variable is the voltage on the capacitor, v. The current that will flow through the resistor is v/R. The equation linking voltage and inflowing current i for a capacitor is

$$i = C\frac{dv}{dt}$$

Since the current in the resistor is flowing *out* of the capacitor, we have

$$i = -\frac{v}{R}$$

so

$$C\frac{dv}{dt} = -\frac{v}{R}$$

or

$$\frac{dv}{dt} = -\frac{v}{RC}$$

3. First let us assume that the bicycle's angle is small, so that its sine can be assumed to be equal to the value of the angle in radians. Now the state variable can be defined as the distance of the back wheel from the line along which the bicycle is being wheeled. If this distance is x and the length between the wheels is L, then a bit of trigonometry shows that for a small angle the component of the velocity of the back wheel perpendicular to the line is Vx/L toward it, where V is the forward speed. We end up with

$$\frac{dx}{dt} = -V\frac{x}{L}$$

4. As the motor speeds up, a backward electromotive force (back-emf) is generated that opposes the applied voltage V. If the angular velocity is ω, the motor current will be proportional to

$$V - k\omega$$

So, we have

$$\frac{d\omega}{dt} = aV - b\omega$$

where $b = ak$. When the motor reaches its top speed, the acceleration will be zero, so

$$\omega_{max} = aV/b$$

Equations of this sort apply to a vast range of situations. A rainwater barrel has a small leak at the bottom. The rate of leakage is proportional to the depth, H, and so

$$\frac{dH}{dt} = -kH$$

The water will leak out until eventually the barrel is empty. But suppose now that there is a steady flow *into* the barrel, sufficient to raise the level (without leak) at a speed u. Then the equation becomes

$$\frac{dH}{dt} = -kH + u$$

What will the level of the water settle down at now? When it has reached a steady level, however long that takes, the rate of change of depth will have fallen to zero, so

$$\frac{dH}{dt} = 0$$

It is not hard to see that $-kH + u$ must also be zero, and so

$$H = u/k$$

Now, if we really want to know the depth as a function of time, a mathematical formula can be found for the solution. But let us try another approach first: simulation. See http://www.EssMech.com/6/2.

6.2 SIMULATION

With very little effort, we can construct a computer program that will imitate the behavior of the barrel. The depth right now is H, and we have already described the rate of change of depth dH/dt as $(-kH + u)$. In a short time dt, the depth will have changed by

$$(-kH + u)dt$$

so that in program terms we have

```
H = H + (-k*H + u)*dt
```

This will work as it stands in most computer languages, although some might insist on it ending with a semicolon. Even when wrapped up in input and output statements to make a complete program, the simulation is very simple. In QBasic it is

```
PRINT "Plot of Leaky Barrel
INPUT "Initial level - 0 to 40 (try 0 first) "; h
INPUT "Input U, 0 to 20      (try 20 first) "; u

k = 0.5
dt = 0.01  'Edit to try various values of steplength dt

Screen 12
WINDOW (-.5, 0)-(5.5, 40)
PSET(t, h)      'starting point
DO
   h = h + (-k * h + u) * dt  'This is the simulation
   t = t + dt
   LINE - (t, h)  'This joins the points with lines
LOOP UNTIL t > 5
```

Take care! This simulation will not be exact. The change in H over time dt will be accurate only if dt is very small. For longer timesteps, dH/dt will change during the interval and the simulated change in H will be in error.

Try values such as $dt = 1$ to see the error. See also that for small values, reducing dt makes no perceptible change.

6.3 SOLVING THE FIRST-ORDER EQUATION

At last we must consider the formal solution of the simple first-order example, where we assume that the system is linear. The treatment here may seem overelaborate, but later on we will apply the same methods to more demanding systems.

By using the variable x instead of H or T_{coffee} or such, we can put all these examples into the same form

$$\frac{dx}{dt} = ax + bu \tag{6.1}$$

where a and b are constants that describe the system. u is an input, which can simply be a constant such as $T_{ambient}$ in the first example or else be a signal that we can vary as a control.

Rearranging, we see that

$$\frac{dx}{dt} - ax = bu$$

Since we have a mixture of x and dx/dt in this expression, we cannot simply integrate it. We must somehow find a function of x and of time that will fit in with both terms on the left of the equation.

If we multiply both sides by a mystery function $f(t)$, we get

$$\frac{dx}{dt} f(t) - axf(t) = -buf(t) \tag{6.2}$$

Now consider

$$\frac{d}{dt}(xf(t))$$

When we differentiate by parts, we see that

$$\frac{d}{dt}(xf(t)) = \frac{dx}{dt} f(t) + xf'(t)$$

where $f'(t)$ is the derivative of $f(t)$.

If we can choose $f(t)$ so that

$$f'(t) = -af(t)$$

then this will fit the lefthand side of Equation (6.2) to give

$$\frac{d}{dt}(xf(t)) = buf(t) \tag{6.3}$$

and we can simply integrate both sides to get the solution.

The function that satisfies

$$f'(t) = -af(t)$$

is

$$f(t) = e^{-at}$$

Now we can integrate both sides of Equation (6.3) to obtain

$$[xe^{-at}]_0^t = \int_0^t bue^{-at}\,dt$$

that is

$$x(t)e^{-at} - x(0) = \int_0^t bue^{-at}\,dt$$

so

$$x(t) = x(0)e^{at} + e^{at}\int_0^t bue^{-at}\,dt$$

Now, if a is positive, the first term will represent a value that will run off to infinity as time increases. If our system is to be stable, a has to be negative. So the coffee cup, the water barrel and the bicycle are stable, but the bank account is not.

If u remains constant throughout the interval 0 to t, we can simplify this still further:

$$x(t) = x(0)e^{at} + ub(e^{at} - 1)/a \tag{6.4}$$

We will come back to this equation when we look at sampled data control.

6.4 SECOND-ORDER PROBLEMS

I hope that you had no difficulty coming to grips with first-order systems, ones that had a single state variable. The following are second-order systems. You should be able to spot two state variables for each of them. You should also be able to write two differential equations for each example:

1. A mass hanging on a spring, bouncing vertically
2. A pendulum swinging in a plane
3. The distance between the back wheel of a bicycle and a straight line when the handlebar angle is varied (small movements away from the line)
4. The voltage on a capacitor when it, a resistor and an inductance are all connected in parallel
5. The position of a servomechanism where acceleration is the input

Once again, give it a try before reading on. Answers are as follows:

1. For the mass, bouncing vertically on a spring, there are two state variables. The first is the height x of the mass above the rest position, and the second is its upward velocity v. The first differential equation can be seen as subtle or obvious, depending on how you look at it

$$\frac{dx}{dt} = v$$

since the rate of change of position is simply the velocity. The second equation is not quite as easy. The rate of change of velocity is the acceleration. Now the acceleration is proportional to the deflection, the displacement away from the rest position, where the constant is the stiffness of the spring divided by the mass. The second equation is therefore

$$\frac{dv}{dt} = -\frac{Sx}{M}$$

If we add an input to the system, by allowing the top of the spring to be moved up or down a distance u, we have

$$\frac{dv}{dt} = \frac{S(u-x)}{M}$$

2. This is almost exactly the same as the previous example. This time the state variables can be taken as the angle of the pendulum and the angle's rate of change. Instead of the constant S/M, however, the constant for the

second equation is now g/L, the acceleration due to gravity divided by the length of the pendulum.

3. Unlike the previous bicycle example, this time both front and back wheels can move away from the straight line. We could take these two distances as the state variables. If we call them x for the rear wheel and w for the front wheel, then we see that the bicycle is pointing at an angle $(w - x)/L$ to the line. The rate of change of the rear-wheel distance will be this angle times the forward speed:

$$\frac{dx}{dt} = (w - x)\frac{V}{L}$$

The direction in which the front wheel is pointing will be $(w - x)/L + u$, where u is the handlebar angle, so

$$\frac{dw}{dt} = -(w - x)\frac{V}{L} + Vu$$

This looks rather different from the other examples. But the choice of state variables is not unique. Instead of the deflection of the front wheel, we could instead have taken the angle the bicycle is pointing as our second state variable. If we call this angle a, we have

$$\frac{dx}{dt} = Va$$

For the second equation, since $a = (w - x)/L$, we have

$$\frac{da}{dt} = \left(\frac{dw}{dt} - \frac{dx}{dt}\right)\Big/L$$

that is

$$\frac{da}{dt} = u\frac{V}{L}$$

4. The two things that cannot change instantaneously are the voltage v on the capacitor and the current i through the inductor. For the inductor, we have

$$L\frac{di}{dt} = v$$

and for the capacitor we have

$$C\frac{dv}{dt} = \text{current into the capacitor}$$

$$= -i - \frac{v}{R}$$

where R is the resistance. Rearrange these slightly, and you arrive at equations for di/dt and dv/dt.

5. This is almost too easy! State variables are now position x and velocity v, so we have

$$\frac{dx}{dt} = v$$

and

$$\frac{dv}{dt} = bu$$

where u is the drive applied to the servomotor and b is a constant.

6.5 MODELING POSITION CONTROL

A servomotor drives a robot axis to position x. The speed of the axis is v. The acceleration is proportional to the drive current u; for now there is no damping.

Can we model the system to deduce its performance?

We have just found equations for the rate of change of x and v:

$$\frac{dx}{dt} = v$$
$$\frac{dv}{dt} = bu$$

We can carry their values forward over an interval dt by adding dt times these rates of change to their values, just as we did for the water barrel.

With a fixed input, the response will not be very interesting. The real use for such a simulation will be to try out various values of feedback. We can start with a position error of 1, say, and ask the user to input values for f, the position feedback, and d, the damping or velocity feedback.

Application of feedback means, "Giving u a value depending on the state." So, before we update the variables, we must make

```
u = -f * x - d * v
```

The following code will perform the simulation; if we define input u in terms of acceleration, rather than motor drive, we can make $b = 1$:

```
SCREEN 12
WINDOW (-.05, -1.1)-(2.05, 1.1)
INPUT "Feedback, damping (suggest 2,2 to start) "; f, d
LINE (0, 0)-(2, 0), 9  'Axis, blue
dt = .0001       'Make dt smaller to slow display
x = 1            'Initial values
v = 0
t = 0
PSET (0, x)        'move to the first point
DO
   u = - f * x - d * v    'u is determined by feedback
   x = x + v * dt         'This is the simulation
   v = v + u * dt
   t = t + dt

   LINE -(t, x)           'This displays the result
LOOP UNTIL t > 2
```

Start with *f*, *d* values of 2, 2.
Next try 10, 5. What do you notice?
How about 1000, 50?
Now try 10000, 200.

It seems that we can speed up the response indefinitely by giving bigger and bigger values of feedback. Is control really so simple?

Of course, the answer is "No." Our simulation has assumed that the system is linear, that doubling the input doubles everything else. But if we have a real motor, there is a limit at which it can accelerate. Let us suppose that the units in our simulation are meters and that it runs for 2 s.

Let us also make the maximum acceleration 10 meters per second. After the u = line, insert two lines

```
IF u > 10 THEN u = 10
IF u < -10 THEN u = -10
```

to impose a limit on u. Now run the program and try all the pairs of values again.

Values 2,2 and 10,5 give the same sluggish responses as before. But 1000,50 overshoots wildly, and 10000,200 is even worse. The limit on the motor drive has had a dramatic effect on performance.

But now try 10000,1600. There is no overshoot and the response has settled in considerably less than one second. It seems that if we know how, we can design good controllers for nonlinear systems. There will be more on that in Chapter 10.

You can borrow a couple of lines from the code in Chapter 3 to construct a real-time "target," changing as you tap the keys, then feed back `(target - x)` in the equation for `u`.

Now you can look at some examples on the book's Website, at http://www.essmech.com/6/5/, to see similar simulations written in JavaScript. Do not just run them; experiment with them and modify their code.

6.6 MATRIX STATE EQUATIONS

We have succeeded in finding a way to simulate the second-order problem, and there seems no reason why the same approach should not work for third, fourth, and more. How can the approach be formalized?

First we must find a set of variables that describe the present state of the system—in this case x and v.

They must all have derivatives that can be expressed as combinations of just the state variables themselves and the inputs, together with constant parameters that are properties of the system.

We have a set of equations that express the change of each variable from instant to instant. If there should happen to be some unknown term, then we have clearly left out one of the state variables; we must hunt for its derivative to work it in as an extra equation.

In the present position control example, the equations can be laid out as follows:

$$\frac{dx}{dt} = v$$

$$\frac{dv}{dt} = bu$$

As soon as a mathematician sees a pair of equations, there is an irresistible urge to put them in the form of a single matrix equation

$$\frac{d}{dt}\begin{bmatrix} x \\ v \end{bmatrix} = \begin{bmatrix} 0 & 1 \\ 0 & 0 \end{bmatrix}\begin{bmatrix} x \\ v \end{bmatrix} + \begin{bmatrix} 0 \\ b \end{bmatrix} u$$

and then to push the shorthand even further. The vector, for that is what x and v have become, is represented by a single symbol $\mathbf{x}$. The 2×2 matrix is given the symbol A, and the matrix that multiplies u is given the symbol B. Just in case u might "fatten up" and have two components, it is also made a vector, $\mathbf{u}$. So we have

$$\frac{d\mathbf{x}}{dt} = A\mathbf{x} + B\mathbf{u}$$

Well, it does look a lot neater. There is one more change. There is a convention to represent the time derivative by a dot over the variable, so we end up with

$$\dot{\mathbf{x}} = A\mathbf{x} + B\mathbf{u} \tag{6.5}$$

which looks very much like the form we used for the first-order systems.

But this still describes the open-loop system, the one that we would like to change with some feedback. How can we deal with feedback in matrix terms? The secret is in the u = line. The mixture of variables that we feed back can be expressed in matrix terms as

$$\mathbf{u} = F\mathbf{x} + G\mathbf{d}$$

where **d** is some external demand such as the target value for position. Now we can substitute into the state equation to get

$$\dot{\mathbf{x}} = A\mathbf{x} + B(F\mathbf{x} + G\mathbf{d})$$

which simplifies to

$$\dot{\mathbf{x}} = (A + BF)\mathbf{x} + BG\mathbf{d}$$

Apart from A having become $A + BF$ and B having become BG, this has exactly the same form as Equation (6.5). The effect of our feedback has been to change the A and B matrices to values that we like better. But how can we decide what we will like?

You will have to wait until Chapter 8, when you will become skilled in the art of eigenvalues.

6.7 ANALOG SIMULATION

It should be obvious by now that our state equations tell us which inputs to apply to integrators that will have outputs corresponding to the state variables' values.

It is ironic that analog simulation "went out of fashion" just as the solid-state operational amplifier was perfected. Previously the integrators had involved a variety of mechanical, hydraulic, and pneumatic contraptions, followed by an assortment of electronics based on magnetic amplifiers or thermionic valves. Valve amplifiers were common even in the late 1960s, and required elaborate stabilization to overcome their drift. Power consumption was high and air-conditioning essential.

Soon an operational amplifier was available on a single chip, then four to a chip at a price of a few cents. But by then it was deemed easier and more accurate to simulate a system on a digital computer. The costly part of analog computing had become that of achieving tolerances of 0.01% for resistors and capacitors, and of constructing and maintaining the large precision patchboards on which each problem was set up.

In the laboratory, the analog computer still has its uses. Leave out the patchboard, and solder up a simple problem directly. Forget the 0.1% components—the parameters of the system being modeled are probably not known to better than a percent or two, anyway. Add a potentiometer or two to set up feedback gains, and a lot of valuable experience can be acquired. Take the problem of the previous section, for example.

We have seen in Chapter 5 that an analog integrator can be made from an operational amplifier, but that the signal integrates in the negative sense when a positive signal is applied.

To produce an output that will change in the positive sense, we must follow this integrator with an inverter. That will mean, too, that with both signs of the signal available, we can attach a potentiometer between them to try both negative and positive feedback.

The circuit shown in Figure 6.1 can give some sort of simulation of the position control problem, although as it stands, the range of gains you can try will be very limited.

One feature it does represent is limits. The amplifiers cannot give voltages outside their supply rails of +12 and −12 V. You will see that the feedback signals have been mixed in an inverter, connected to the first integrator with a 10 kΩ resistor. This gives an effective gain of 10, and the effect on the amplifier limit is equivalent to saying that the motor is capable of accelerating at a rate of 10 m/s. This gain of 10 will apply to the feedback coefficients, but they will still be much smaller than the values you used in the "professional" position control experiment.

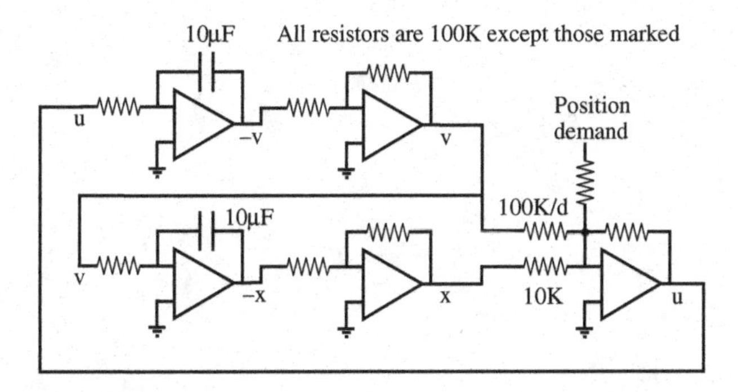

Figure 6.1 *Simulation circuit, gain of 10 from mixer.*

6.8 MORE FORMAL COMPUTER SIMULATION

The simulations we have seen so far are "run up on the spot" as simply and concisely as possible. For more general use, however, we need a more formal methodology.

Software can be written in a host of languages, including QBasic, Visual Basic, JavaScript, or even a package such as Matlab, but the simulation will have a common structure:

1. Define constants and variables.
2. Set variables to their initial conditions, and define the timestep.
3. Begin the loop.
4. Calculate the drive input(s)
5. Calculate the rates of change of the state variables, using the state equations.
6. Update the state variables, by adding rate of change *dt.
7. Update the current time, by adding dt to it.
8. Plot the variables, or capture them for plotting later
9. Repeat the loop until the end of the simulation.

The state equations do not have to be linear. They can include limits or geometric functions as necessary, depending on the detail that we are trying to achieve. We can simulate continuous control, with timesteps that are small in order to preserve accuracy. We can simulate discrete-time control where the drive is allowed to change only at intervals of many steps of the continuous system's update.

The essential requirement is that the state equations used must be an accurate representation of the system.

7

Vectors, Matrices, and Tensors

For both state space control theory and kinematics, we can take advantage of matrix methods.

There is a tendency among mathematicians to regard matrices as arcane and mystic entities, with cryptic properties that reward a lifetime of study. Engineers can be duped into this point of view if they are not careful.

7.1 MEET THE MATRIX

Matrices are, in fact, just a form of shorthand that can come in very useful when a lot of calculating operations are involved. There are strict rules to observe, but when used properly matrices, vectors, and tensors are mere tools that are the servant of the engineer.

You will probably have first encountered matrices in the solution of simultaneous equations. To take a simple example, the equations

$$5x + 7y = 2$$

$$2x + 3y = 1$$

can be "tidied up" by separating the coefficients from the variables in the form

---

Essentials of Mechatronics, by John Billingsley

$$\begin{bmatrix} 5 & 7 \\ 2 & 3 \end{bmatrix} \begin{bmatrix} x \\ y \end{bmatrix} = \begin{bmatrix} 2 \\ 1 \end{bmatrix}$$

where the variables x and y are now conveniently grouped as a vector. Now the multiplication rule has defined itself.

We move across the top row of the matrix, multiplying each element by the corresponding component as we move down the vector to its right, adding up these products as we go. We put the resulting total in the top element, here $5x + 7y$.

Then we do the same for the next row, and so on.

7.2 MORE ON VECTORS

What does a *vector* actually "mean"? The answer has to be "anything you like." Anything, that is, that cannot be represented by a single number but requires a string of numbers to define it. It could even be a shopping list:

$$5 \text{ oranges} + 3 \text{ lemons} + 2 \text{ grapefruit}$$

can be written in matrix format as

$$[\text{orange} \quad \text{lemon} \quad \text{grapefruit}] \begin{bmatrix} 5 \\ 3 \\ 2 \end{bmatrix}$$

which we might write in a line of text as (orange, lemon, grapefruit) (5,3,2)′ or else place the dot between them that we use for *scalar product*. The numbers on the right have defined a "mixture" of the items on the left.

Rather than fruit, we are more likely to apply vectors to coordinate systems—but we are still just picking from a list.

We might define $\boldsymbol{i}, \boldsymbol{j}$, and $\boldsymbol{k}$ to be *unit vectors* all at right angles, say, east, north, and up. We can call them *basis vectors*.

When we say that point P has coordinates (2,3,4)′, we mean that to get there, you start at the origin and go 2 m east, then 3 m north, and 4 m up.

We could write this as

$$2\boldsymbol{i} + 3\boldsymbol{j} + 4\boldsymbol{k}$$

which is a mixture of the basis vectors defined by a matrix multiplication—vectors are just skinny matrices.

Now, when we turn our minds to applications, we can see many uses for vector operations. When a force $\boldsymbol{F}$ moves a load a distance $\boldsymbol{x}$, the work done is given by their scalar product $\boldsymbol{F} \cdot \boldsymbol{x}$.

As before, we take products of corresponding elements and add them up, to get a scalar number.

We usually think in terms of "the matrix multiplies the vector." But how about thinking of the vector multiplying the matrix? What does it do to it? Consider the following matrix:

$$\begin{bmatrix} 1 & 2 & 3 \\ 4 & 5 & 6 \\ 7 & 8 & 9 \end{bmatrix}\begin{bmatrix} x \\ y \\ z \end{bmatrix} = \begin{bmatrix} 1x+2y+3z \\ 4x+5y+6z \\ 7x+8y+9z \end{bmatrix}$$

From one perspective, the top element is equal to the scalar product of the top row of the matrix with the vector $(x,y,z)'$. Similarly, the other elements are the scalar products of the vector with the middle and bottom rows of the matrix, respectively.

So we have

> The product of a matrix and a (column) vector is made up of the scalar products of the vector with each of the rows of the matrix.

But there is another way of seeing it. The answer is the same as

$$\begin{bmatrix} 1 \\ 4 \\ 7 \end{bmatrix} x + \begin{bmatrix} 2 \\ 5 \\ 8 \end{bmatrix} y + \begin{bmatrix} 3 \\ 6 \\ 9 \end{bmatrix} z$$

So we also have

> The product of a matrix and a column vector is a mixture of the vectors that make up the columns of the matrix.

Suppose that point P is defined in terms of a second set of basis vectors, $\boldsymbol{u}$, $\boldsymbol{v}$, and $\boldsymbol{w}$, so that its coordinates $(x,y,z)'$ mean $x\boldsymbol{u} + y\boldsymbol{v} + z\boldsymbol{w}$. To find the coordinates in terms of $\boldsymbol{i}$, $\boldsymbol{j}$, and $\boldsymbol{k}$, we simply multiply and add up the contributions from $\boldsymbol{u}$, $\boldsymbol{v}$, and $\boldsymbol{w}$.

We can "transform the coordinates" by multiplying $(x,y,z)'$ by a matrix made up of columns representing vectors $\boldsymbol{u}$, $\boldsymbol{v}$, and $\boldsymbol{w}$, to end up with a vector for P as a mixture of $\boldsymbol{i}$, $\boldsymbol{j}$, and $\boldsymbol{k}$.

7.3 MATRIX MULTIPLICATION

Often we will find a need to multiply one matrix by another. To see this in action, let us look at another simple "mixing" example.

In a candy store, "scrunches", "munches," and "chews" are on sale.

Also on sale are "Jumbo" bags each containing 2 scrunches, 3 munches, and 4 chews, and "Giant" bags containing 5 scrunches, 6 munches and only one chew. If I purchase 7 Jumbo bags and 8 Giant bags, how many of each sweet have I bought?

The bag contents can be expressed algebraically as

$$J = 2s + 3m + 4c$$

and

$$G = 5s + 6m + 1c$$

or in matrix form as

$$\begin{bmatrix} J \\ G \end{bmatrix} = \begin{bmatrix} 2 & 3 & 4 \\ 5 & 6 & 1 \end{bmatrix} \begin{bmatrix} s \\ m \\ c \end{bmatrix}$$

Note that matrices do not have to be square, as long as the terms to be multiplied correspond in number.

Now my purchase of 7 Jumbo bags and 8 Giant bags can be written as

$$7J + 8G$$

or in grander form as the product of a row vector with a column vector:

$$\begin{bmatrix} 7 & 8 \end{bmatrix} \begin{bmatrix} J \\ G \end{bmatrix}$$

But I can substitute for the J,G vector to obtain

$$\begin{bmatrix} 7 & 8 \end{bmatrix} \begin{bmatrix} 2 & 3 & 4 \\ 5 & 6 & 1 \end{bmatrix} \begin{bmatrix} s \\ m \\ c \end{bmatrix}$$

To get numerical counts of scrunches, munches, and chews we have to calculate the product of a numerical row vector with a numerical matrix. As before, we march across the row(s) of the one on the left, taking the scalar product with the columns on the right.

The answer is what common sense would give.

From 7 Jumbo bags, with scrunches at 2 to a bag, we find 7 times 2 scrunches.
From 8 Giant bags, we find 8 times 5 more, giving a grand total of 54.

The final answer is

$$[54 \quad 69 \quad 36]\begin{bmatrix} s \\ m \\ c \end{bmatrix}$$

that is 54 scrunches, 69 munches, and 36 chews.

Now the shop is selling an Easter bundle of 3 Jumbo bags and a Giant bag, and still has in stock Christmas bundles of 2 Jumbo bags and 4 Giant bags. If I buy five Easter packs and one Christmas pack, how many scrunches, munches, and chews will I have?

As an exercise, write down the matrices involved and multiply them out by the rules that we have found. (Your answer should be 89 scrunches + 105 munches + 77 chews.)

The mathematician will still worry about the order in which the matrix multiplication is carried out. We must not alter the order of the matrices, but we can group the pairs for calculation in either of two ways.

The Christmas and Easter bags can first be opened to reveal a total of Jumbo and Giant bags, then these can be expanded into individual sweets. Alternatively, work out the total of each sweet for a Christmas bag and for an Easter bag first. The result must be the same. (Check it.)

Mathematicians would say that "multiplication of matrices is associative:"

$$ABC = (AB)C = A(BC)$$

7.4 TRANSPOSITION OF MATRICES

Our mixed fruit multiplication can be written as

$$[\text{orange} \quad \text{lemon} \quad \text{grapefruit}]\begin{bmatrix} 5 \\ 3 \\ 2 \end{bmatrix}$$

or equally well as

$$[5 \quad 3 \quad 2]\begin{bmatrix} \text{orange} \\ \text{lemon} \\ \text{grapefruit} \end{bmatrix}$$

giving 5 oranges + 3 lemons + 2 grapefruit in both cases—this result is in the form of a scalar. But note that in reversing the order in which we multiply the vectors, we have had to transpose them.

Transposing a scalar is not very spectacular—but when two matrices are multiplied together to give another matrix, $C = AB$, then, if we wish to find out the transpose of C, we must both transpose A and B and reverse the order in which we multiply them:

$$C' = B'A'$$

7.5 THE UNIT MATRIX

One last point to note before moving on is that

$$\begin{bmatrix} 1 & 0 & 0 \\ 0 & 1 & 0 \\ 0 & 0 & 1 \end{bmatrix}\begin{bmatrix} x \\ y \\ z \end{bmatrix} = \begin{bmatrix} x \\ y \\ z \end{bmatrix}$$

The matrix with 1s down its diagonal and 0s elsewhere has the special property that its product with any vector or matrix leaves that vector or matrix unchanged. Of course, there is not just one unit matrix; they come in all sizes to fit the rows of the matrix that they have to multiply. This one is the 3×3 version.

7.6 COORDINATE TRANSFORMATIONS

It has been mentioned that vector geometry is usually introduced with the aid of three orthogonal unit vectors: $\mathbf{i}, \mathbf{j}$, and $\mathbf{k}$.

For now, let us keep to two dimensions and consider just $(x,y)'$, meaning $x\mathbf{i} + y\mathbf{j}$.

Now suppose that there are two sets of axes in action. With respect to our first set the point is $(x,y)'$ but with respect to a second set it is $(u,v)'$. Just how can these two vectors be related?

What we have in effect is one pair of unit vectors $\mathbf{i}, \mathbf{j}$, and another pair, $\mathbf{l}$, $\mathbf{m}$, say. Since both sets of coordinates represent the same vector, we have

$$\mathbf{x}\boldsymbol{i} + y\boldsymbol{j} = u\boldsymbol{l} + v\boldsymbol{m}$$

Now each of the vectors $\boldsymbol{l}$ and $\boldsymbol{m}$ must be expressible in terms of $\boldsymbol{i}$ and $\boldsymbol{j}$. Suppose that

$$\boldsymbol{l} = a\boldsymbol{i} + b\boldsymbol{j}$$

$$\boldsymbol{m} = c\boldsymbol{i} + d\boldsymbol{j}$$

or in matrix form

$$[\boldsymbol{l} \quad \boldsymbol{m}] = [\boldsymbol{i} \quad \boldsymbol{j}]\begin{bmatrix} a & c \\ b & d \end{bmatrix}$$

We want the relationship in this slightly twisted form, because we want to substitute into

$$[\boldsymbol{l} \quad \boldsymbol{m}]\begin{bmatrix} u \\ v \end{bmatrix}$$

to eliminate vectors $\boldsymbol{l}$ and $\boldsymbol{m}$ to get

$$[\boldsymbol{i} \quad \boldsymbol{j}]\begin{bmatrix} a & c \\ b & c \end{bmatrix}\begin{bmatrix} u \\ v \end{bmatrix}$$

Now the ingredients must match:

$$\begin{bmatrix} x \\ y \end{bmatrix} = \begin{bmatrix} a & c \\ b & d \end{bmatrix}\begin{bmatrix} u \\ v \end{bmatrix}$$

Although this exercise is now graced with the name "vector geometry," we are merely adding up mixtures in just the same form as the antics in the candy store.

To convert our $(u,v)'$ coordinates into the $(x,y)'$ frame, we simply multiply the coordinates by an appropriate matrix that defines the mixture.

Suppose, however, that we are presented with the values of x and y and are asked to find $(u,v)'$. We are left trying to solve two simultaneous equations:

$$x = au + cv$$

$$y = bu + dv$$

In traditional style, we multiply the top equation by d and subtract c times the second equation to obtain

$$dx - cy = (ad - bc)u$$

and in a similar way, we find

$$-bx + ay = (ad - bc)v$$

which we can rearrange as

$$\begin{bmatrix} u \\ v \end{bmatrix} = \frac{1}{(ad - bc)} \begin{bmatrix} d & -c \\ -b & a \end{bmatrix} \begin{bmatrix} x \\ y \end{bmatrix}$$

where the constant $1/(ad - bc)$ multiplies each of the coefficients inside the matrix.

If the original relationship between $(x,y)'$ and $(u,v)'$ was

$$\begin{bmatrix} x \\ y \end{bmatrix} = T \begin{bmatrix} u \\ v \end{bmatrix}$$

then we have found an "inverse matrix" such that

$$\begin{bmatrix} u \\ v \end{bmatrix} = T^{-1} \begin{bmatrix} x \\ y \end{bmatrix}$$

The value of $(ad - bc)$ obviously has special importance—we will have great trouble in finding an inverse if $(ad - bc) = 0$. Its value is the "determinant" of the matrix T.

7.7 MATRICES, NOTATION, AND COMPUTING

In a computer program, rather than using separate variables x, y, u, v, and so on, it is more convenient mathematically to use "subscripted variables" as the elements of a vector.

The entire vector is then represented by the single symbol $\boldsymbol{x}$, which is made up of several elements x_1, x_2, and so on.

Matrices are now made up of elements with two suffices:

$$A = \begin{bmatrix} a_{11} & a_{12} & a_{13} \\ a_{21} & a_{22} & a_{23} \\ a_{31} & a_{32} & a_{33} \end{bmatrix}$$

In a computer program, the subscripts appear in brackets, so that a vector could be represented by the elements **`X(1)`**, **`X(2)`**, and **`X(3)`**, while the elements of the matrix are **`A(1,1)`**, **`A(1,2)`**, and so on.

It is in matrix operations that this notation really earns its keep. Suppose that we have a relationship

$$x = Tu$$

where the vectors have three elements and the matrix is 3×3. Instead of a massive block of arithmetic, the entire product is expressed in just five lines of Basic program:

```
FOR I=1 TO 3
   X(I)=0
   FOR J=1 TO 3
      X(I)=X(I)+T(I, J)*U(J)
   NEXT J
NEXT I
```

For the matrix product $C = AB$, the program is hardly any more complex:

```
FOR I=1 TO 3
   FOR J=1 TO 3
      C(I,J)=0
      FOR K=1 TO 3
         C(I,J)=C(I,J)+A(I,K)*B(K,J)
      NEXT K
   NEXT J
NEXT I
```

Or in Java or C it becomes

```
for(i = 1; i<=3; i++){
   for(j = 1; j<=3; j++) {
      c[i][j] = 0;
      for(k = 1; k<=3;k++) {
            c[i][j] += a[i][k]*b[k][j];
      }
   }
}
```

These examples would look almost identical in a variety of languages and would show the same economy of programming effort.

In Matlab the shorthand of matrix operations goes even farther—but there is a danger that the engineroom will be lost to view behind the paintwork.

> Clearly, if we are to try to analyze any except the simplest of systems by computer, we should first represent the problem in a matrix form.

But beware!!

If you have no computer to hand, it will almost certainly be quicker, easier, and less prone to errors to use non-matrix methods to solve the problem.

7.8 EIGENVECTORS

If we multiply a vector and a matrix, what do we get?

We get another vector. For example

$$\begin{bmatrix} 1 & 2 \\ -1 & 4 \end{bmatrix}\begin{bmatrix} 1 \\ 0 \end{bmatrix} = \begin{bmatrix} 1 \\ -1 \end{bmatrix}$$

From the vector $(1,0)'$, we get $(1,-1)'$. This new vector is not only a different "size"; it represents a different direction. Another example is:

$$\begin{bmatrix} 1 & 2 \\ -1 & 4 \end{bmatrix}\begin{bmatrix} 0 \\ 1 \end{bmatrix} = \begin{bmatrix} 2 \\ 4 \end{bmatrix}$$

So, from the vector $(0,1)'$, we get $(2,4)'$—again in a new direction.

Are there any vectors that can be multiplied by the matrix

$$\begin{bmatrix} 1 & 2 \\ -1 & 4 \end{bmatrix}$$

to give another vector in the same direction?

If we start with $(x,y)'$, another vector in the same direction will be $(\lambda x, \lambda y)'$—where λ is some constant.

We are looking for a vector $\boldsymbol{x}$ for which

$$\boldsymbol{Ax} = \lambda \boldsymbol{x}$$

or

$$\boldsymbol{Ax} = \lambda \boldsymbol{Ix}$$

where I is the unit matrix. We can move both terms to the lefthand side to get

$$\boldsymbol{Ax} - \lambda \boldsymbol{Ix} = \boldsymbol{0}$$

or

$$(A - \lambda I)\boldsymbol{x} = \mathbf{0}$$

where the $\mathbf{0}$ is a vector with all components zero.

You will recall that we could consider the matrix–vector product as a mixture of the columns of the matrix.

So here, if the vector $\boldsymbol{x}$ is not $\mathbf{0}$, we have a combination of the columns of $(A - \lambda I)$ that will give $(0,0)'$.

Remember also that to evaluate a determinant of a matrix, you can first add multiples of columns to other columns of the matrix without changing the determinant's value.

Thus we have a way to reduce a column of $(A - \lambda I)$ to all zeros, and so its determinant must be zero.

Now, when we construct $A - \lambda I$ and take its determinant, we get

$$\det\begin{vmatrix} 1-\lambda & 2 \\ -1 & 4-\lambda \end{vmatrix} = 0$$

which we can expand as

$$(1-\lambda)(4-\lambda) - (-1)2 = 0$$

or

$$\lambda^2 - 5\lambda + 6 = 0$$

So, we have not just one value for λ, but two: 2 and 3.

If we substitute the value 2, we get

$$\begin{bmatrix} 1-2 & 2 \\ -1 & 4-2 \end{bmatrix}\begin{bmatrix} x \\ y \end{bmatrix} = \begin{bmatrix} 0 \\ 0 \end{bmatrix}$$

which is satisfied if $\boldsymbol{x} = (2,1)'$.

Let us try it out:

$$\begin{bmatrix} 1 & 2 \\ -1 & 4 \end{bmatrix}\begin{bmatrix} 2 \\ 1 \end{bmatrix} = \begin{bmatrix} 4 \\ 2 \end{bmatrix}$$

So $A\boldsymbol{x} = 2\boldsymbol{x}$, just as we hoped to find, and $\boldsymbol{x}$ is an *eigenvector* of A. The value of λ is called an *eigenvalue.*

As an exercise, find the other eigenvector, corresponding to eigenvalue $\lambda = 3$.

If the matrix A is $n \times n$, the equation for λ will be nth order and there will be n roots. But the method is just the same:

1. Write down $(A - \lambda I)$ and take its determinant.
2. Equate the determinant to 0, giving a polynomial for λ.
3. Solve this, to get a set of n eigenvalues.
4. For each eigenvalue, substitute that value back into $(A - \lambda I)\boldsymbol{x} = \mathbf{0}$, getting a set of simultaneous equations for the elements of $\boldsymbol{x}$.
5. Solve these equations, and you have each corresponding eigenvector.

8

Mathematics for Control

8.1 DIFFERENTIAL EQUATIONS

8.1.1 Breaking Down the State Equations

In Section 6.6, we saw how a system could be described by a matrix state equation of the form

$$\dot{\mathbf{x}} = A\mathbf{x} + B\mathbf{u}$$

in which there are several simultaneous first-order equations.

We have looked at an example where

$$\dot{x} = v$$
$$\dot{v} = bu$$

and we could consider applying feedback

$$u = (-6x - 5v)/b$$

to get

$$\dot{v} = -6x - 5v$$

Essentials of Mechatronics, by John Billingsley

In matrix form these equations are

$$\begin{bmatrix} \dot{x} \\ \dot{v} \end{bmatrix} = \begin{bmatrix} 0 & 1 \\ -6 & -5 \end{bmatrix} \begin{bmatrix} x \\ v \end{bmatrix} \tag{8.1}$$

We can eliminate v from the two equations to get the "conventional" form of a single second-order differential equation

$$\ddot{x} = -6x - 5\dot{x}$$

or

$$\ddot{x} + 5\dot{x} + 6x = 0$$

8.1.2 Solving the Single-Variable Equation

There are a number of ways to solve such an equation. The high-school approach is to say try e^{mt}. If

$$x = e^{mt}$$

then

$$\dot{x} = me^{mt}$$

and

$$\ddot{x} = m^2 e^{mt}$$

so

$$\ddot{x} + 5\dot{x} + 6x = (m^2 + 5m + 6)e^{mt}$$

The exponential will be nonzero for all finite values of m and t, so we equate the quadratic to zero and solve it, in this case getting roots $m = -2$ and $m = -3$.

The general solution will be

$$x = Ae^{-2t} + Be^{-3t}$$

where A and B are constants determined by the initial conditions

$$x(0) = A + B$$
$$\dot{x}(0) = -2A - 3B$$

8.1.3 Solving the Matrix Equation Directly

Now let us consider the matrix form again. Can we solve Equation (8.1) in a more direct way?

The equation has the form

$$\dot{\mathbf{x}} = A\mathbf{x}$$

Suppose that $\mathbf{x}$ happens to be in the direction of one of the eigenvectors of A, which we will call $\boldsymbol{\xi}$. We could write

$$\mathbf{x} = n\boldsymbol{\xi}$$
$$\dot{\mathbf{x}} = \dot{n}\boldsymbol{\xi}$$

where $\boldsymbol{\xi}$ is a constant vector. Now $A\boldsymbol{\xi} = \lambda\boldsymbol{\xi}$, since that is how eigenvectors and eigenvalues are defined, so $A\mathbf{x}$ will be in the same direction as $\mathbf{x}$;

$$\dot{\mathbf{x}} = A\mathbf{x}$$

tells us that

$$\dot{n}\boldsymbol{\xi} = \lambda n\boldsymbol{\xi}$$

and since $\boldsymbol{\xi}$ is constant, we have

$$\dot{n} = \lambda n$$

which has the solution

$$n = n(0)e^{\lambda t}$$

If λ is positive, this will represent a function that will keep growing to infinity. If λ is negative, it will die away to zero. For stability, this λ must be negative.

But there will be as many eigenvalues and eigenvectors as the order of the system. For second- and higher-order systems, we can express $\mathbf{x}$ as a mixture of the eigenvectors. So now we need *all* the eigenvalues to be negative, since if any one of them should be positive, the corresponding component will grow to infinity.

However, some of the roots could be complex.

Now

$$e^{j\omega t} = \cos(\omega t) + j\sin(\omega t)$$

So

$$e^{(\lambda+j\omega)t} = e^{\lambda t}e^{j\omega t}$$
$$= e^{\lambda t}(\cos(\omega t) + j\sin(\omega t))$$

If the real part of the root is positive, the response will be a sine wave that keeps on growing. So, for stability, the real parts of every one of the roots must be negative.

Let us take another look at the response of the position control system, by finding the eigenvalues of the matrix that describes it:

$$\begin{bmatrix} 0 & 1 \\ -6 & -5 \end{bmatrix}$$

We take the determinant of $A - \lambda I$

$$\begin{vmatrix} -\lambda & 1 \\ -6 & -5-\lambda \end{vmatrix}$$

and arrive at the quadratic equation

$$\lambda^2 + 5\lambda + 6 = 0$$

The roots are $\lambda = -2$ and $\lambda = -3$. Does that sound familiar?

Replace the m in Section 8.1.2 by s, and you will see something resembling the notation of the Laplace transform. Of course, the roots are the same, yet again.

8.2 THE LAPLACE TRANSFORM

The mathematical justification of the Laplace transform involves integrals over infinite time. The inverse requires an infinite contour integral in the complex frequency domain. But all of this is irrelevant to the way the notation is used by a mechatronic engineer.

8.2.1 The Basis of the Transform

The significant property of its definition is that the transform of the derivative of a function is the variable s times the transform of the function, minus the value of the function at $t = 0$:

$$\mathcal{L}(\dot{x}) = s\mathcal{L}(x) - x(0)$$

This achieves two things. It eliminates derivatives, turning each differentiation into a variable s. It also gives a formal method of dealing with the initial conditions. The result is a function of s for which the corresponding function of time can be looked up in a table. In effect, the table of transforms is a cook book full of "Here's one I prepared earlier."

Now, when we take the transform of our equation for the position system, we get

$$\mathcal{L}(\ddot{x}+5\dot{x}+6x)=(s^2+5s+6)\,\mathcal{L}(x)-sx(0)-\dot{x}(0)-5x(0)$$

so, here is that quadratic again!

With no other input, this expression is equal to zero, so we can rearrange it to get

$$\mathcal{L}(x)=\frac{(s+5)x(0)+\dot{x}(0)}{s^2+5x+6}$$

So now we know the Laplace transform of x, but what is it as a function of time?

The cornerstone of the method is the *uniqueness theorem*, which states that there is one and only one function of time that corresponds to any transform in s. If we have constructed a table of functions and their transforms, then, if we can match the transform, we have found the right function.

In the case above, we can factorize the denominator and split the expression into partial fractions. If, for example, $x(0) = 2$ and $\dot{x}(0) = -5$, we get

$$\mathcal{L}(x)=\frac{1}{s+2}+\frac{1}{s+3}$$

and of course when we look them up in the table, we find the same pair of exponentials as before.

In the mid-1950s, before the Laplace notation became fashionable, the Heaviside D operator was used for the same purpose. Where today we see polynomials in s, then we would have seen polynomials in D, although an extra s appears in the denominators of the functions in the table of transforms.

In the D operator notation, the transform that is just 1 corresponds to the unit step, which is zero for all negative time and has value 1 for all positive time. The Laplace function $1/s$ corresponds to the unit step, but the inverse of the Laplace function 1 is the unit impulse. This has a time integral of 1, but is infinitesimally thin, so that it has to be infinitely tall. It is not a very comfortable function to have to deal with.

8.2.2 Transfer Functions

A useful application of the Laplace transform notation is for the expression of transfer functions. They have an important place in the analysis of control systems, as long as they are not held to be the one and only method.

Consider yet again the motor system described by the equation $\ddot{x} = u$, and suppose yet again that we wish to apply feedback. This time, however, we have no tacho signal and have only x to feed back.

We know that making $u = -ax$ will give us

$$\ddot{x} = -ax$$

which is the equation for simple harmonic motion. Undamped oscillation is not the best kind of control that we might hope for. So, as we did in the experiment of Chapter 3, we try to "guess" the velocity from x.

To estimate the velocity, we first construct x_{slow}, where

$$\frac{d}{dt}x_{\text{slow}} = k(x - x_{\text{slow}})$$

In Laplace terms, ignoring initial conditions, this becomes

$$sX_{\text{slow}} = k(X - X_{\text{slow}})$$

where capitals are used for the transforms, so

$$(s+k)X_{\text{slow}} = kX$$

or

$$X_{\text{slow}} = \frac{k}{s+k}X$$

We estimated the velocity as $k(x - x_{\text{slow}})$ so that

$$V_{\text{est}} = kX - k\frac{k}{s+k}X$$

$$= \frac{sk}{s+k}X$$

So, now that we have an estimated velocity to feed back, let us try

$$\ddot{x} = -ax - bv_{\text{est}}$$

which in Laplace terms is expressed as

$$s^2X = -aX - b\frac{sk}{s+k}X$$

We can multiply through by $(s + k)$ and reorganize to get

$$(s^3 + ks^2 + (a+bk)s + ak)X = 0$$

To test stability, we look at the roots of the cubic in s. The response will involve terms in e^{st} for each root of the polynomial. As before, if the real part of any root is positive, the exponential will run away and the system will clearly be unstable. So once again we see that all the roots must have negative real parts.

Lemma. A cubic can always has one real root, so it can be factorized into the form

$$\begin{aligned}&(s+p)(s^2+qs+r)\\&= s^3+(p+q)s^2+(pq+r)s+pr\end{aligned}$$

Now we know that if and only if p, q, and r are positive, the roots will have negative real parts and the system will be stable. An easy deduction is that the three coefficients of the polynomial must be positive, but there is another condition. Look at the product of the middle two coefficients

$$(p+r)(pq+r)$$

If p, q, and r are positive, this is clearly greater than pr, which is just one of terms when expanded. But this is the product of the first and last coefficients. So, for stability, the product of the middle two coefficients must be greater than the product of the outer two.

In the example above, if a, b, and k are all positive, we can see that the condition for stability is satisfied. So here is a theory confirming that estimating the velocity by this method will always work as far as stability is concerned, but we have to look deeper to select values for the "best performance."

8.2.3 Transfer Functions and Matrices

We can mix the transform method with the matrix state equations, too. When we have

$$\dot{\mathbf{x}} = A\mathbf{x} + B\mathbf{u}$$

we can take the transform to get

$$s\mathbf{X} = A\mathbf{X} + B\mathbf{U}$$

which we can rearrange by introducing a unit matrix I, to get

$$(sI - A)\mathbf{X} = B\mathbf{U}$$

from which we get

$$\mathbf{X} = (sI - A)^{-1} B\mathbf{U}$$

This gives us a *transfer function matrix* that enables us to express each element of **X** in terms of the elements of input function **U**.

8.3 DIFFERENCE EQUATIONS

Until now, when we have used the computer to update the state variables, we have been careful to make the timestep small, so that the approximation to continuous differential equations will be sufficiently accurate. But can we find another way to analyze the system that recognizes the discrete-time nature of computer control?

8.3.1 Sequences of Discrete-Time Samples

As far as the computer is concerned, x is not a continuous function but is defined by a sequence of sampled values, $x_0, x_1, x_2, x_3. \ldots$ The analysis is made much easier if we assume that these are taken at regular equal intervals of time, T, so that our continuous and discrete systems are linked by

$$x_n = x(nT)$$

The computer outputs its control variable u_n very shortly after the measurement of x_n, and u remains constant until the next sample time.

With continuous variables, we defined our equations in terms of rate of change. Now we can instead look at the difference between samples, so that instead of

$$\frac{dx}{dt} = ax + bu$$

we have something like

$$x_{n+1} - x_n = cx_n + du_n$$

but it is all so much simpler if instead of differences we just think of the next value.

$$x_{n+1} = (1+c)x_n + du_n$$

At the end of Section 6.3, we constructed a solution to the differential equation by multiplying both sides by an exponential and integrating. We got Equation (6.4)

$$x(t) = x(0)e^{at} + ub(e^{at} - 1)/a$$

which calculated the value of x an interval t after applying a constant input u. If the interval is T, we have

$$x(T) = x(0)e^{aT} + u(0)b(e^{aT} - 1)/a$$

With slight modification this will tell us the value of x at time $(n + 1)T$ in terms x and input u at time nT

$$x((n+1)T) = x(nT)e^{aT} + u(nT)b(e^{aT} - 1)/a$$

or in terms of our sequence of samples

$$x_{n+1} = x_n e^{aT} + u_n b(e^{aT} - 1)/a$$

This is in a form similar to that of our original state equation, showing that the next x is a linear combination of the present state and the present input. We could write this as

$$x_{n+1} = ax_n + bu_n$$

but we would risk confusion between the continuous and discrete parameters.

Let us settle for

$$x_{n+1} = px_n + pu_n$$

where

$$p = e^{aT}$$

and

$$q = b(e^{aT} - 1)/a$$

Now if the input is zero, we obtain

$$x_{n+1} = px_n$$

so

$$x_1 = px_0$$
$$x_2 = p^2 x_0$$
$$x_n = p^n x_0$$

For stability, p^n must not grow indefinitely, so the magnitude of p must not be greater than unity. For a disturbance to decay to zero, we require that

$$|p| < 1$$

8.3.2 Discrete-Time State Equations

We have found a solution for the first-order case, but what if the system is of higher order? Can we use similar methods to solve the matrix differential equation? Can we use

$$\dot{\mathbf{x}} = A\mathbf{x} + B\mathbf{u}$$

to get a discrete-time form?

In Section 6.3, we multiplied both sides by e^{-at} to get an expression that we could integrate. But is there such a thing as e^{-At} when A is a matrix?

We can expand e^{-at} as an infinite series

$$e^{-at} = 1 - at + a^2t^2/2! - a^3t^3/3!\ldots$$

and when we differentiate it term by term, we see a result that is $-a$ times the series with which we started.

In the same way, we can define

$$e^{-At} = I - At + A^2t^2/2! - A^3t^3/3!\ldots$$

and by differentiating term by term, then comparing powers of t against the original series, we see that its derivative is $-e^{-At}A$. So now

$$\frac{d}{dt}\left(e^{-At}\mathbf{x}\right) = e^{-At}\dot{\mathbf{x}} - e^{-At}A\mathbf{x}$$

and by an integral similar to that in Section 6.3, we arrive at

$$\mathbf{x}(t) = e^{At}\mathbf{x}(0) + (e^{At} - I)A^{-1}B\mathbf{u}$$

when $\mathbf{u}$ is constant over the interval. Hence

$$\mathbf{x}_{n+1} = e^{At}\mathbf{x}_n + (e^{At} - I)A^{-1}B\mathbf{u}_n$$

which we can write as

$$\mathbf{x}_{n+1} = P\mathbf{x}_n + Q\mathbf{u}_n$$

The matrix P is the *state transition matrix*, sometimes written as $\Phi(T)$.

With zero input, the state is multiplied by P between samples, so that

$$\mathbf{x}_n = P^n\mathbf{x}_0$$

If λ is an eigenvalue of P, and if $\mathbf{x}_0$ is the corresponding eigenvector, then

$$\mathbf{x}_n = \lambda^n\mathbf{x}_0$$

so if the magnitude of any eigenvalue is greater than unity, the state will run off to infinity. For a disturbance to decay to zero, we require that

$$|\lambda| < 1$$

for every eigenvalue of P.

8.3.3 A Shortcut to Discrete-Time State Equations

For a system like the position controller, there is a more direct way to get the discrete-time state equations. We merely solve the equations in a direct way.

We have

$$\ddot{x} = bu$$

so

$$\dot{x}(t) = \dot{x}(0) + but$$
$$x(t) = x(0) + \dot{x}(0)t + but^2/2$$

We can rewrite these, giving values of x and v at time T, as

$$x(T) = x(0) + v(0)t + ubt^2/2$$
$$v(T) = v(0) + ubT$$

The state equation is therefore

$$\begin{bmatrix} x_{n+1} \\ v_{n+1} \end{bmatrix} = \begin{bmatrix} 1 & T \\ 0 & 1 \end{bmatrix} \begin{bmatrix} x_n \\ v_n \end{bmatrix} + \begin{bmatrix} bT^2/2 \\ bT \end{bmatrix} u_n$$

8.4 THE *Z* TRANSFORM

A mathematician can make the Laplace transform look simple in comparison with the z transform. With contour integrals in the complex frequency plane, summation of infinite series, and an explanation in terms of trains of impulses, the subject can be made somewhat forbidding.

8.4.1 The "Next" Operator

There is, of course, another way to look at the topic. While the Laplace s can be seen as shorthand for d/dt, z can be regarded as meaning "next."

The discrete-time matrix state equation is

$$\mathbf{x}_{n+1} = P\mathbf{x}_n + Q\mathrm{u}_n$$

which we can regard as defining "next" $\mathbf{x}$. For the transform, we can write

$$z\mathbf{X} = P\mathbf{X} + Q\mathbf{U}$$

and get a discrete transfer function in the form

$$\mathbf{X} = (zI - P)^{-1} Q\mathbf{U}$$

It is easy to make a connection between the z operator and lines of software. When a variable is changed, we can regard the assignment statement as setting the "next" value.

So, from

```
xslow = xslow + k * (x - xslow) * dt
```

we can replace dt by T and get

$$\mathbf{next}(x_{\text{slow}}) = x_{\text{slow}} + kT(x - x_{\text{slow}})$$

or in transform terms

$$zX_{\text{slow}} = X_{\text{slow}} + kT(X - X_{\text{slow}})$$

so

$$(z-1+kT)X_{\text{slow}} = kTX$$

This gives us the discrete transfer function

$$X_{\text{slow}} = \frac{kT}{z-(1-kT)}X$$

Now V_{est} was given by

$$V_{\text{est}} = k(X - X_{\text{slow}})$$

(there is no extra z because this is "algebra" rather than a state equation)

$$V_{\text{est}} = k\left(1 - \frac{kT}{z-(1-kT)}\right)X$$

$$= k\frac{z-1}{z-(1-kT)}X$$

To work out the transfer function of the double integrator, we look at the final state equation in the previous section:

$$\begin{bmatrix} x_{n+1} \\ v_{n+1} \end{bmatrix} = \begin{bmatrix} 1 & T \\ 0 & 1 \end{bmatrix}\begin{bmatrix} x_n \\ v_n \end{bmatrix} + \begin{bmatrix} bT^2/2 \\ bT \end{bmatrix}u_n$$

We can write

$$(z-1)X = TV + b(T^2/2)U$$
$$(z-1)V = bTU$$

so, substituting for V and dividing through by $(z-1)$, we obtain

$$X = \frac{z+1}{(z-1)^2}\frac{bT^2}{2}U$$

Now, if

$$U = -fX - dV_{\text{est}}$$

we can substitute for V_{est} to obtain a polynomial in z multiplying X. The roots of this polynomial will determine whether the system is stable.

As an exercise, try the algebra and see what you can tell about f, d, and kT for stability. There are more conditions to satisfy than in the continuous case.

You can also try *pole assignment*, where you choose three roots that you would like and manipulate the values of f, d, and kT to match the equation coefficients. Try matching three equal roots of 0.5.

The solution is as follows:

$$X = \frac{z+1}{(z-1)^2}\frac{bT^2}{2}U$$
$$= \frac{z+1}{(z-1)^2}\frac{bT^2}{2}\left(-fX - dk\frac{z-1}{z-(1-kT)}X\right)$$

So, multiplying through by the denominators, we have

$$2(z-1)^2(z-(1-kT))X = -f(z-(1-kT))X - dk(z-1)X$$

or bringing everything to the left and taking out the factor X, we obtain

$$\left\{2(z-1)^2(z-(1-kT)) + f(z-(1-kT)) + dk(z-1)\right\}X = 0$$

We end up inspecting the roots of

$$2z^3 - (6-2kT)z^2 + (6-4kT+f+dk)z - (f+dk-fkT) = 0$$

Remember that we are not looking for the simple condition that all the roots have negative real parts, but instead we must show that their magnitudes should all be less than unity.

Instead of struggling, we can "cheat" by saying that we would like three equal roots of 0.5; in other words, the polynomial in z is equivalent to

$$2(z-0.5)^3 = 0$$

(The 2 is there to make the coefficients of z^3 match.)

$$2z^3 - 3z^2 + 1.5z - 0.25 = 0$$

By equating coefficients, we have

$$2kT = 3, \text{ so } kT = 1.5$$
$$f + dk = 1.5$$
$$f + dk - 1.5f = 0.25$$

thus

$$1.5f = 1.25$$

giving

$$f = \frac{5}{6} \quad \text{and} \quad dk = \frac{2}{3}$$

A free decision can still be made concerning the sampling interval.

Do not forget that the values of 0.5 have been pulled out of thin air, without any real justification. The actual behavior of the system might be better assessed by simulation.

8.5 CONVOLUTION AND CORRELATION

Although these seem to be rather abstruse mathematical tricks, heaped with double-summation sigma signs, they are remarkably useful.

8.5.1 Convolution

Having just come to grips with discrete-time control and the z transform, it is appropriate to deal with convolution first.

Let us apply a time function $u(nT)$ to our system. If we wish, we can think of this as a train of outputs to a digital-to-analog converter—there is no need to get tied up with impulses.

Suppose first that we apply just one output, of value 1 at $n = 0$ and zeros from then on. We can express this as a sequence (1,0,0,0, . . .).

How should we describe the output? We are interested only in the sample values at $t = 0$, $t = T$, $t = 2T$, and so on, which we can write as $y(nT)$ or y_n. We might have measured the sequence of values in an experiment or deduced the function from mathematical manipulation of state equations, it does not matter.

So, if we apply the input sequence

$$(1, 0, 0, 0 \ldots)$$

to our system we have a special *unit response*:

$$(h_1, h_2, h_3, h_4, \ldots, h_n, \ldots)$$

If the first input is of size u_0 instead of 1, we will have an output at each value of nT:

$$y_n = u_0 h_n$$

Now suppose instead that we apply an input at $t = T$, so that

$$u = (0, u_1, 0, 0, 0 \ldots)$$

Everything will happen one sample later, so that the output at $(n + 1)T$ is

$$y_{n+1} = u_1 h_n$$

so

$$y_n = u_1 h_{n-1}$$

In the first case the result of the input had time nT to "mature," but the second input a sample later has only had time $(n - 1)T$ to mature until we sample the output at time nT.

We can go on considering the effect of each individual input u_i at time iT, which will be

$$y_n = u_i h_{n-i}$$

but when we have to consider the effect of the whole input sequence combined, we must add them all up—assuming that the system is linear.

So, now we have an expression with a summation

$$y_n = \sum u_i h_{n-i}$$

Over what range do we have to perform the summation?

Well, it is no use starting before $i = 0$, since we assume that the input sequence started only then. There is no point in continuing beyond $i = n$, unless our system is able to respond to inputs that will happen in the future. (Since we might not always be dealing with time functions, this could sometimes be the case.)

The mathematician would say that y is obtained from the *convolution* of u with h.

In some cases we can regard our system as a filter, which we apply to process the sequence u. It might, for example, be a smoothing filter to present weather data or gasoline prices more neatly. In the cases where we have to perform the summation all the way from the start, it would be called an *infinite impulse response* (IIR) filter, meaning that the effect of a single input will take forever to die away.

But we can use other filters with a limited "window." We might just want to take the average of the latest 10 values, in which case we start summing only from $i = n - 9$. Alternatively, we may consider our lowpass filter to "run out of steam" after the unit response has had 10 intervals to decay, so that we chop off the sequence at that point to save computing effort. In either case, we will call the filter a *finite impulse response* (FIR) filter.

We will later see this sort of convolution in action in image processing.

8.5.2 Correlation

In convolution, we multiply the terms of one sequence taken left to right by terms from another taken right to left and add up the result.

Correlation is very similar, except the terms are taken in the same direction, but with some displacement between them. So what is it useful for?

Global Positioning System (GPS) satellites transmit a "song" consisting of a repeated *pseudorandom binary sequence* (PRBS). We can think of this as a sequence of +1s and –1s like this:

```
+ + + + − − − + − − + + − + −
```

The 15 symbols repeat to give

```
+ + + + − − − + − − + + − + − + + + + − − − + − − +
```

Now, if we multiply each symbol by itself, we will, of course, get a string of +1s, and if we sum these over a cycle, we will get the answer 15. But what happens if we move the first sequence—let us call it the *template*—relative to the second that we can consider a test sequence. First let us move it by just one symbol:

```
  + + + + − − − + − − + + − + −             template
+ + + + − − − + − − + + − + − + + + + − −   test
  + + + − + + − − + − + − − − −             product
```

and now when you sum the terms in the product, you get the answer –1. In fact, you will get this answer when you shift the template relative to the test sequence by any number of symbols except an exact cycle of 15.

Of course, GPS uses much longer sequences, 1023 for *coarse acquisition* and a huge number for the precision signal. However, the principle is the same. By correlating the received signal against a sequence generated in the receiver, it is possible to get a measure of the delay time from satellite to ground—and hence the distance. In fact, each satellite generates a different sequence, and the correlation of one satellite's "song" against another is always near zero.

(If you are interested, the sequence above is $a_3 \oplus a_4$—the next value is 1 if the third to the left is different from the fourth to the left and –1 if they are the same.)

So, we have an expression for the correlation

$$C(n) = \sum a_i b_{i+n}$$

where we sum over the range of the template, a, to give an answer that is a function of the shift, n.

The uses are endless. We can examine an audio signal to try to recognize particular sounds in it. By launching into two dimensions and a double summation, we can examine image data to look for specific objects or characters.

Image correlation is something not to be entered lightly, though. If our template is just 32 pixels square, we have over 1000 multiplications and additions for a single point of the result. But if the test image is 320×320 pixels, we can consider 288 values of shift in each direction. We arrive at some 80 million operations to process a single image, and that is for a specific size and orientation of the template.

We have to take a little care with the template, which will probably contain analog values rather than simple 1s and –1s. We must reduce its mean to zero, so that we will not get a significant response when it is correlated against a constant. We must also smooth its ends, so that the "chopped off" data at the limits of summation do not look like anything of interest.

9

Robotics, Dynamics, and Kinematics

After all the electronic sensing, signal processing, and computing have been put into effect, most applications must result in some mechanical movement. We might be required to look at the theory of coordinating the axes of a robot to put the workpiece in the correct position or more simply to choose a motor and gearbox to move a load at a safe top speed.

9.1 GEARS, MOTORS, AND MECHANISMS

Electricity is powerful stuff. It is quite easy to relate electrical power to mechanical power in metric units, although pounds and feet will require a lot of conversion factors.

Consider the following:

1 kilogram force = 9.81 newtons
(so a force of one newton is about the weight of an apple)
1 joule = 1 newton - meter
1 watt = 1 joule per second

So a one-watt motor, if it were 100% efficient, could lift a one-kilogram mass at a rate of 10 centimeters per second. To lift a 75-kilogram passenger in an elevator at one meter per second will require

$$75 \times 9.81 \times 1\,\mathrm{W}$$

Now the motor may be only 50% efficient, so provision must be made for 1.5 kilowatts per passenger, plus a lot more for accelerating the cage.

In selecting a motor for a mechatronic task, it is important to allow for sufficient power. But it is also important not to provide excessive power, force, or speed.

Some time in the 1950s, the autopilot of a passenger aircraft decided that the aircraft should plunge vertically. Not surprisingly, the pilot disagreed, but could not disengage the autopilot. The resulting tug-of-war came to an end when the geartrain of the autopilot broke. Ever since, autopilots have been designed with a *shear link*, a sort of mechanical fuse, so that the possible disaster can ultimately be blamed on pilot error.

9.1.1 Calculating Motor Performance

A typical small motor might have a top speed of some 6000–12,000 revolutions per minute, that is, 100–200 revolutions per second. How can we convert this rotary motion into a linear motion of, say, 1 meter per second?

A pulley to match this speed would have to have an effective circumference of between 5 and 10 millimeters—much smaller than practical. With a reduction gear of ratio 30:1, however, the pulley could be between 50 and 100 mm in diameter (remember that π is involved).

There are a number of parameters that will define the motor: the resistance, the stall torque, the no-load speed, the moment of inertia, the rated voltage, and the rated power. We should also consider the starting torque.

When the motor rotates, it generates a back-emf—indeed, any good motor can be used as a generator. There is an important coefficient that we will call k_V, where, if we neglect the starting torque

$$k_V = V_{\text{rated}} 60/(2\pi \,\text{rpm}_{\text{no load}})$$

The generated voltage will be

$$V_{\text{gen}} = k_V \omega$$

where ω is the angular velocity of the rotation. You will note that k_V has been calculated to make the generated voltage equal to the rated voltage at the no-load speed.

A good permanent-magnet DC motor will have a small starting torque and corresponding small starting voltage. If allowed to run freely, it will take little current since it will run at such a speed that the generated back-emf almost equals the supply voltage. If a load is applied to the motor, it will slow down, the back-emf will drop, and the current will increase accordingly until the drive torque is equal to the load torque. That leads us to another important parameter, k_T, such that

$$\text{Torque} = k_T i$$

where i is the current in the motor. We can calculate k_T from the resistance R and the rated stall torque by

$$k_T = \text{stall torque } R/V_{\text{rated}}$$

Under load and at steady speed, the output power is the product of the torque and the angular velocity, so it is given by

$$\begin{aligned} & k_T i\omega \\ & = k_T\omega(V - k_V\omega)/R \end{aligned}$$

When the motor runs free, the output power is zero; when the motor is stalled, the output power is also zero. Maximum mechanical power is obtained at half the no-load speed, when the back-emf V_{gen} will be $V/2$.

At any speed, the power dissipated in the motor as heat is $i^2 R$, while the power taken from the supply is iV. But

$$V = V_{\text{gen}} + iR$$

so the mechanical power output, equal to supply power minus dissipation, is

$$\begin{aligned} & = V_{\text{gen}}\, i \\ & = k_V \omega\, i \end{aligned}$$

But above we saw that this power was

$$k_T \omega\, i$$

so

$$k_T = k_V$$

So we can simply call these two parameters, which are actually the same parameter, k. The voltage generated per radian per second is equal to the meter-newtons of torque per ampere of current.

We also see that at half no-load speed, the output mechanical power is equal to the dissipated heat.

9.1.2 The Effect of an Inertial Load

Now we can set up a differential equation for the motor, when driven with no load from voltage V:

$$I\frac{d\omega}{dt} = ki$$

$$= k\frac{V - k\omega}{R}$$

The motor will accelerate up to its steady speed with time constant $I\,R/k^2$.

When we add an inertial load of mass M, it will increase the effective moment of inertia to $I + M\,r^2$, where r is the "effective pulley," the distance moved by the mass for each radian of motor rotation. This takes any gearbox into consideration.

The maximum acceleration from rest is

$$r\frac{d\omega}{dt}$$

$$= r\frac{kV/R}{I + Mr^2}$$

which will be greatest if the motor, gearbox, pulley, and mass are "matched" so that

$$Mr^2 = I$$

Of course, maximizing the acceleration may not be the most important objective. There may be a standing force on the mass, for example, if the mass moves vertically or if it is part of a machine with a cutting force. If the motor must withstand a disturbance torque at rest, the power taken from the supply will correspond to that torque acting at the motor's top speed. And all that power will be dissipated as heat in the motor.

It may, therefore, be desirable to increase the gear ratio, thereby decreasing the effective pulley, to obtain a compromise between standing torque and peak acceleration. If the gear ratio is doubled, for example, the standing torque is halved while the peak acceleration is reduced from its optimum by only 20%.

The gear ratio can be multiplied by 3.7 before the peak acceleration is halved, although this will reduce the top speed by the same factor of 3.7. Good design is always a matter of compromise.

9.1.3 Mechanisms

When we wish to convert the rotation of a motor to the motion of a load, a pulley is merely one very simple example of a mechanism to use. The choice will often have very little to do with dynamics.

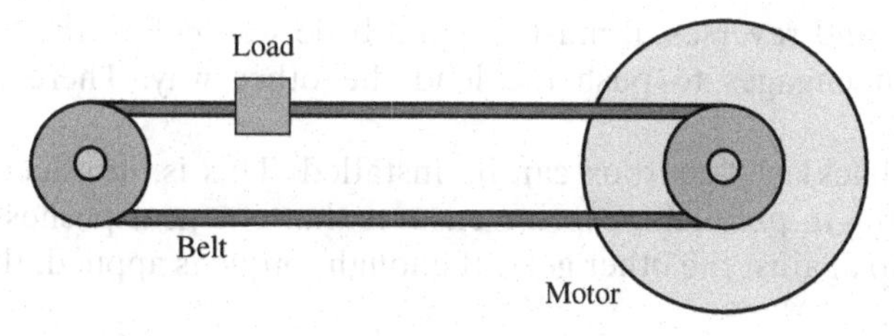

Figure 9.1 Pulley and belt.

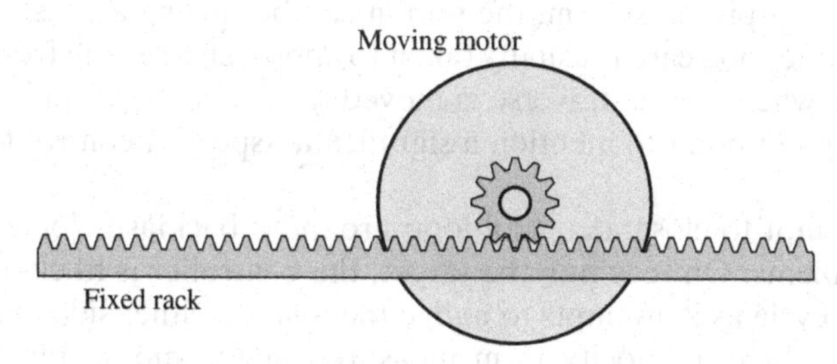

Figure 9.2 Rack and pinion.

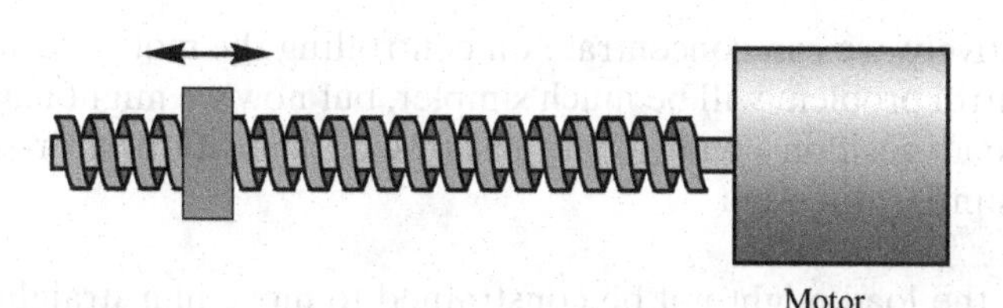

Figure 9.3 Lead screw.

A pulley-and-belt system (Fig. 9.1) has the advantage of simplicity, but has other drawbacks. In its simple form there is the risk of slip, so that there is an error between motion at the motor and motion of the load. This can be avoided with a "toothed belt"—although there is still the issue of stretching of the belt.

A more robust mechanism might appear to be the rack-and-pinion system (Fig. 9.2). A gear on the motor or its gearbox now runs on a rack, or linear gear, running the length of the travel required. This has some penalties of cost, but a greater drawback is that the motor now travels with the mass as part of the load.

Machine tools favor the lead screw (Fig. 9.3), a rod with square-cut threads running the length of the slideway. On one hand, mechanical efficiency is poor; on the other hand, it is insensitive to disturbing forces. It is also likely to suffer from "backlash."

In any system with "teeth," particularly a gearbox, the problem of backlash requires attention. As the motor rotates, the load is pushed along. When the motor stops and reverses, it must rotate a little way before the "other side" of the tooth engages to push the load the other way. There are several remedies.

An "antibacklash" gearbox can be installed. This is, in effect, two gearboxes working in parallel. A spring ensures that one gear pushes the output shaft hard up against the other gear. If enough torque is applied, the backlash is still there.

The same sort of effect occurs if the axis is vertical, so that the gearbox "holds the load up" and contact is always made on the same face of the gear.

In a rack-and-pinion system, the pinion can be sprung against the rack.

In a machine tool, care is usually taken to approach a setting from the same direction, as when a lathe traverse is moved to take a deeper cut.

This is a good point to mention a significant aspect of control theory.

> If we attempt to close a control loop around a backlash element, we will have problems. On reaching the target, the controller is likely to oscillate in a limit cycle as it attempts to nudge the load on either side of zero error. We can include a velocity term measured at the motor, but this might merely convert the dithering to a slower twitch.
>
> Alternatively, we can concentrate on controlling the motor position alone. The control problem will be much simpler, but now we might have an error in the load position equal to the backlash. Elasticity in a drive belt can pose a similar dilemma.

Of course, the load might not be constrained to move in a straight line. The whole appeal of the revolute robot is that arms rotate about pivots at the joints, where any straight lines are the result of cunning coordination of axes working in unison. Other devices rely on mechanisms such as the four-bar linkage that can result in rotation about a "virtual pivot."

Gearbox design is an art in itself. As well as conventional gears, there are worm drives, harmonic drives, sun and planet mechanisms, and many more. When the relationship between motor speed and load speed is to be nonlinear, there are solutions that include elliptic gears. Yet, however complicated the mechanism, we can apply the principle of virtual work.

The product of load force multiplied by the distance that it moves must, ignoring friction losses, be equal to the product of motor torque multiplied by the angle through which it rotates.

9.2 THREE-DIMENSIONAL MOTION

A point P in space is defined by a three-dimensional vector, but the method employed to represent it is not unique. The most obvious form is Cartesian (Fig. 9.4a), in which, the three coordinates are found by resolving the vector

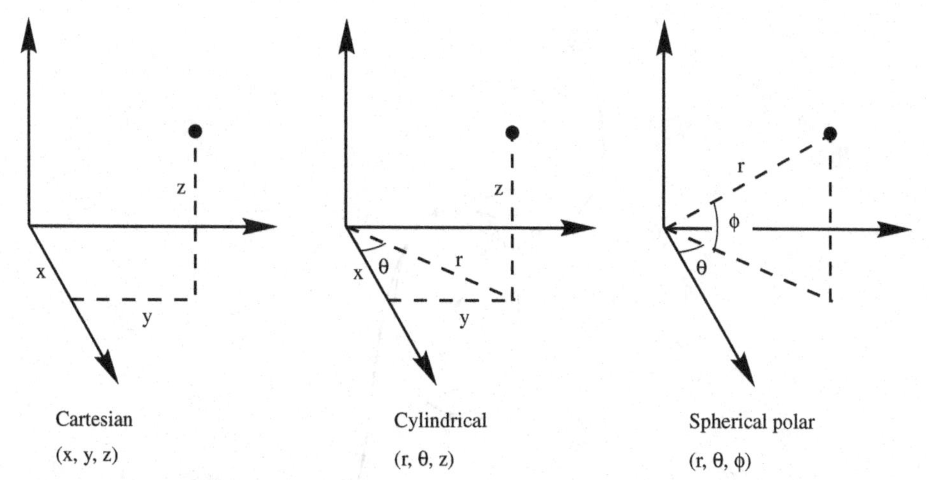

Figure 9.4 *(a) Cartesian (x,y,z), (b) cylindrical (r,θ,z), and (c) spherical polar (r,θ,φ) coordinates.*

from the origin in the directions of three orthogonal vectors through that origin. There are also spherical polar (Fig. 9.4c) coordinates, equivalent to defining the latitude, longitude, and distance of the point from the origin, also cylindrical polar (Fig. 9.4b), in which the point is represented by radius, direction, and height.

Not only is the location of the origin a matter of choice, we can orient the orthogonal vectors of Cartesian coordinates with 3 more degrees of freedom.

For now, however, let us take it that the origin is fixed and that we have three unit vectors $\boldsymbol{i}, \boldsymbol{j}$, and $\boldsymbol{k}$ defining the x, y, and z directions.

As we saw in Chapter 7, our point P can be represented as $(x,y,z)'$, meaning

$$x\boldsymbol{i} + y\boldsymbol{j} + z\boldsymbol{k}$$

When the point moves, x, y, and z will vary as functions of time. Now we can take the derivatives of the vector components to calculate the velocity and acceleration. It is worth making a few remarks about these.

As it moves, P will follow a curve in space (see Fig. 9.5). The velocity vector will be a tangent to this curve at P. The acceleration can be broken into two perpendicular components. One of these is in the same direction as the velocity, representing a change in speed, while the other is perpendicular to the path, aligned through the instantaneous center of rotation, the center of curvature of the path at that point.

This may seem too simple—and it is. When we start to analyze the motion of a robot, we must deal with six dimensions, not just three. We are concerned with solid bodies, not mere points in space. We have three dimensions of

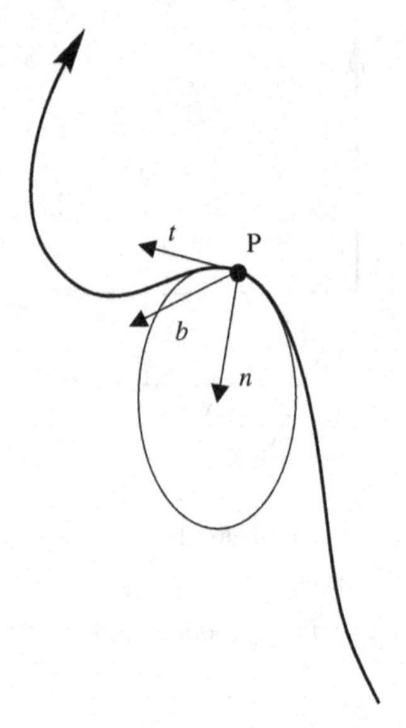

Figure 9.5 Center of curvature.

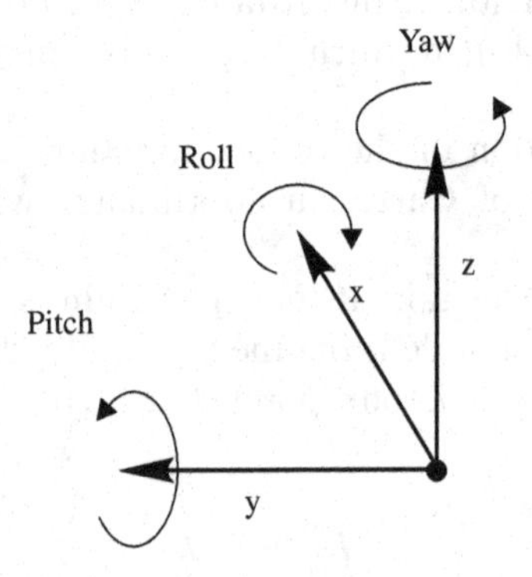

Figure 9.6 Schematic representation of 6 degrees of freedom.

freedom in the location of one particular point of the object, but then we can perform three rotations to orient the object in space. We might think of these rotations as movement about the pitch, roll, and yaw axes of an aircraft (see Fig. 9.6).

Instead of the vector coordinates of just one of its points, we have to think of the position and orientation of the object as being defined by the transformation that maps each of its points to the new position that it takes up. Let us first consider the transformation of rotation.

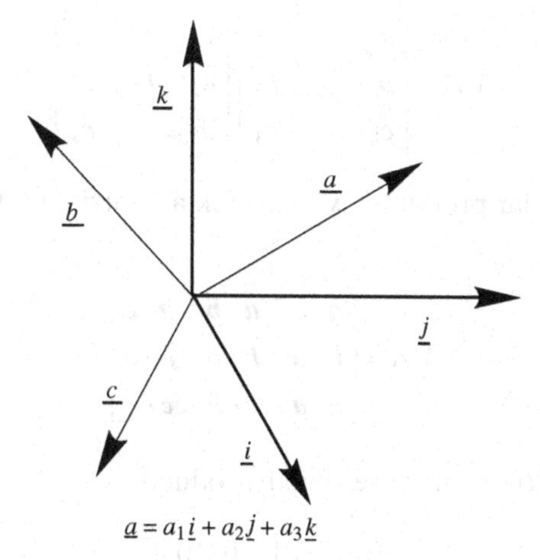

Figure 9.7 *Unit vectors.*

9.2.1 Rotations

We can set up a coordinate system of three orthogonal axes in the object. To start with, these will coincide with our "reference system" axes ***i***, ***j*** and ***k*** that stay fixed. But as we rotate the object about the origin, its axes will move to be three other orthogonal vectors through the origin.

Let us consider three such unit vectors ***a***, ***b***, and ***c***, passing through the origin of our coordinate system and orthogonal to each other (Fig. 9.7).

A point expressed in terms of these vectors as coordinates $(x,y,z)'$ will be

$$\boldsymbol{a}x + \boldsymbol{b}y + \boldsymbol{c}z$$

This can be expanded as

$$[\boldsymbol{i} \quad \boldsymbol{j} \quad \boldsymbol{k}]\begin{bmatrix} a_1 & b_1 & c_1 \\ a_2 & b_2 & c_2 \\ a_3 & b_3 & c_3 \end{bmatrix}\begin{bmatrix} x \\ y \\ z \end{bmatrix}$$

to give the coordinates of the same point in terms of the reference system.

We transform the coordinates to the reference axes by multiplying $(x,y,z)'$ by this matrix A. So, let us look at some of the properties of A.

Since they are unit vectors, $\boldsymbol{a} \cdot \boldsymbol{a} = 1$, $\boldsymbol{b} \cdot \boldsymbol{b} = 1$ and $\boldsymbol{c} \cdot \boldsymbol{c} = 1$. Also, since the vectors are orthogonal, the scalar product of any two different vectors is zero, for example, $\boldsymbol{a} \cdot \boldsymbol{b} = 0$.

Let us consider the product of A with its transpose:

$$A'A = \begin{bmatrix} a_1 & a_2 & a_3 \\ b_1 & b_2 & b_3 \\ c_1 & c_2 & c_3 \end{bmatrix} \begin{bmatrix} a_1 & b_1 & c_1 \\ a_2 & b_2 & c_2 \\ a_3 & b_3 & c_3 \end{bmatrix}$$

Remember the "scalar products" way to look at matrix multiplication. We see that

$$A'A = \begin{bmatrix} \boldsymbol{a}\cdot\boldsymbol{a} & \boldsymbol{a}\cdot\boldsymbol{b} & \boldsymbol{a}\cdot\boldsymbol{c} \\ \boldsymbol{b}\cdot\boldsymbol{a} & \boldsymbol{b}\cdot\boldsymbol{b} & \boldsymbol{b}\cdot\boldsymbol{c} \\ \boldsymbol{c}\cdot\boldsymbol{a} & \boldsymbol{c}\cdot\boldsymbol{b} & \boldsymbol{c}\cdot\boldsymbol{c} \end{bmatrix}$$

But from what we know of these scalar products

$$A'A = \begin{bmatrix} 1 & 0 & 0 \\ 0 & 1 & 0 \\ 0 & 0 & 1 \end{bmatrix}$$

So

$$A'A = I$$

or

$$A' = A^{-1}$$

The rotation transformation matrix is extremely easy to invert!

There is a further property that we have to preserve; the axes must make up a "righthanded" set. The conventional set of axes will be $\boldsymbol{i}$ and $\boldsymbol{j}$, as we draw x and y on a horizontal sheet of graph paper, and $\boldsymbol{k}$ vertically upward in the z direction.

Because it reverses the x coordinate, the matrix

$$\begin{bmatrix} -1 & 0 & 0 \\ 0 & 1 & 0 \\ 0 & 0 & 1 \end{bmatrix}$$

would map a lefthanded glove into a righthanded glove, something no rotation could do. Yet it satisfies the property of having three mutually orthogonal unit vectors as its rows and its columns. What is wrong?

A property of a rotation is that there is an axis about which the rotation takes place. Now, if a vector $\boldsymbol{\xi}$ is aligned with this axis, it is not changed by being transformed by A; in other words

$$A\boldsymbol{\xi} = \boldsymbol{\xi}$$

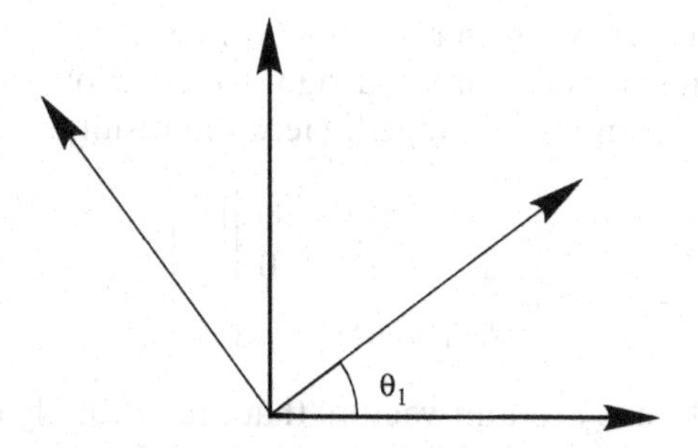

Figure 9.8 Rotation about z axis.

So $\boldsymbol{\xi}$ is an eigenvector of A, with eigenvalue 1. All the eigenvalues of a rotation must be 1, so the determinant of A must be 1. The determinant of the glove-bending matrix is –1, so it cannot represent a rotation.

We should look at some examples of rotation matrices. If we rotate the x–y plane by an angle θ_1 about the z axis (Fig. 9.8), we get new coordinates:

$$(x\cos\theta_1 - y\sin\theta_1,\ x\sin\theta_1 + y\cos\theta_1,\ z)$$

The z component stays the same.

In matrix terms, the transformation is

$$\begin{bmatrix} \cos\theta_1 & -\sin\theta_1 & 0 \\ \sin\theta_1 & \cos\theta_1 & 0 \\ 0 & 0 & 1 \end{bmatrix}$$

A rotation θ_2 about the y axis would be represented by

$$\begin{bmatrix} \cos\theta_2 & 0 & \sin\theta_2 \\ 0 & 1 & 0 \\ -\sin\theta_2 & 0 & \cos\theta_2 \end{bmatrix}$$

Note that a positive rotation is "clockwise looking out along the axis," so this tips the x axis downward.

If we multiply the matrices together to get the result of applying both transformations, we will start to build up a string of sines and cosines that will be lengthy to write and muddling to read. Therefore, we use considerable abbreviation and write $\cos\theta_1$ as c_1, $\sin\theta_1$ as s_1, and so on. If we apply these in order, the transformed coordinates will be

$$\begin{bmatrix} c_2 & 0 & s_2 \\ 0 & 1 & 0 \\ -s_2 & 0 & c_2 \end{bmatrix}\begin{bmatrix} c_1 & -s_1 & 0 \\ s_1 & c_1 & 0 \\ 0 & 0 & 1 \end{bmatrix}\begin{bmatrix} x \\ y \\ z \end{bmatrix}$$

Note that the transformation that is applied first is closest to the vector; in other words, the matrices are ordered right to left. Note, too, that the order is important and must not be changed. Here the result is

$$\begin{bmatrix} c_1 c_2 & -s_1 c_2 & s_2 \\ s_1 & c_1 & 0 \\ -s_2 c_1 & s_1 s_2 & c_2 \end{bmatrix} \begin{bmatrix} x \\ y \\ z \end{bmatrix}$$

Check that the columns are unit vectors that are mutually orthogonal.

See rotations in action at www.essmech.com/9/2/1.htm

So far we have been considering transformations that leave the origin fixed, but we must also be able to move the coordinates anywhere in three dimensions.

9.2.2 Translations

To move an object a vector distance, we simply add that vector to every one of its points.

For example, a point $(x,y,z)'$ can be moved a vector distance $(1,2,3)'$ to arrive at $(x + 1, y + 2, z + 3)'$—it's not really difficult! To find the new vector, we simply add the displacement to it.

The problem is that we now have two different processes for dealing with the two types of movement: rotation and translation. One involves multiplying the point coordinates by a 3×3 matrix, while the other involves adding constants to each component. Can we find some way of gluing them together into a single operation? If we can, we can start to deal with combinations of transformations, such as "screwing" where the object is rotated at the same time as it is moved along the rotation axis.

We have to appease the mathematicians! Rotation is a transformation given by a simple multiplication of a vector by a matrix, but the ability to add a constant to the result requires an *affine* transformation.

However, there is a way around the problem. Suppose that instead of writing our vector as $(x,y,z)'$ we write it as $(x,y,z,1)'$.

What is the 1 for? It gives something for a matrix to grab onto to add a translation d to the vector! But now the vector has four components, and the matrix is 4×4.

We can "partition" a matrix to see its various parts in action, so if we write $T\boldsymbol{x}$ for the product of our point with a transformation matrix, now 4×4, we can break it down as follows.

$$\begin{bmatrix} A & \boldsymbol{d} \\ 000 & 1 \end{bmatrix} \begin{bmatrix} \boldsymbol{x} \\ 1 \end{bmatrix} = \begin{bmatrix} A\boldsymbol{x} + \boldsymbol{d} \\ 1 \end{bmatrix}$$

Thus, at the expense of changing our matrices to 4×4, where the bottom row is always $(0,0,0,1)$, we can apply any combination of rotations and translations, just by multiplying the T matrices together.

This transformation is called the Denavit–Hartenberg (or D–H) matrix.

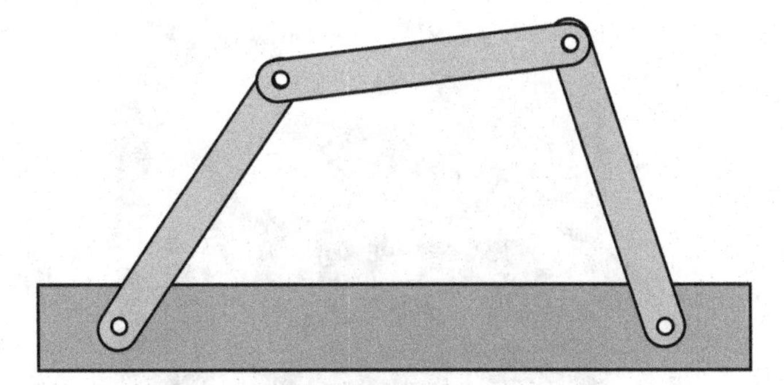

Figure 9.9 *Four-bar linkage (sometimes called three-bar).*

9.3 KINEMATIC CHAINS

The most usual form for a robot is a chain of links with actuated joints between them. These joints can be *revolute*, a sort of powered hinge, or *prismatic*, with one member sliding past another. We will refer to both types as *axes*. Although some kinematic chains can be "closed," such as the four-bar linkage of Figure 9.9, most robots are "open" where only one end of the chain is fixed.

When we consider the toolpiece of a robot, its location in space has been transformed by the motion of every axis in turn that moves it. Before we can address the task of deciding on joint angles or displacements to put the tool where we want it, we have to derive an expression for its location and orientation in terms of the joint axis variables.

9.3.1 Chains of Axes

When we have just one movable axis, there is a single transformation and all is straightforward. When we have a robot such as the *Unimation Puma*, with 6 degrees of freedom, we have to be systematic in the way that we analyze it.

Let us start with $\boldsymbol{i}, \boldsymbol{j}$, and $\boldsymbol{k}$ as the usual x,y,z axes fixed in the mounting of the robot and call them *frame 0*. We need to know the transformation that will convert the coordinates of anything held in the gripper into coordinates with respect to the reference frame 0 in the robot's base.

We can define a succession of frames as we make our way along the robot to the gripper. Each of these frames will have a local x, y, and z direction related by some transformation to the next frame. Some transformations will relate to the variable angles that make up the axes; others will simply take us from one end to the other of a link such as the "forearm."

We can choose the frames so that the transformations between them are extremely simple, involving either a rotation about one of the axes or a translation along one of the axes.

Let us see this in action (Fig. 9.10).

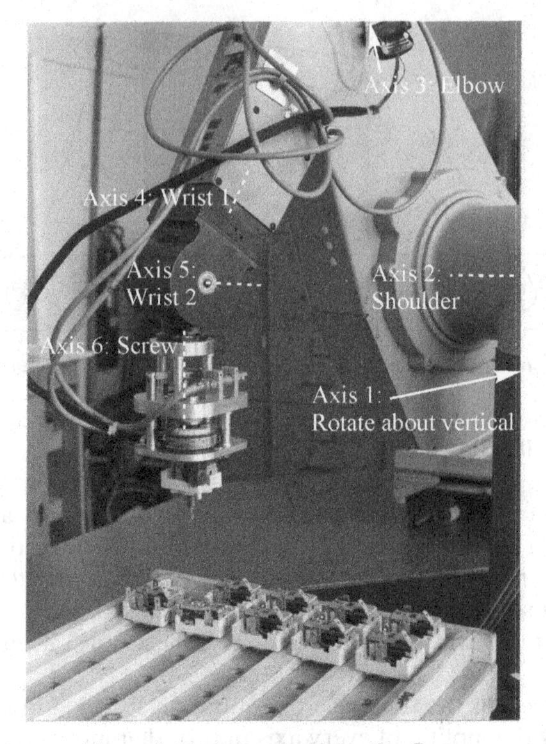

Figure 9.10 *Axes of a Unimation Puma.*

The joints of the Puma can be thought of as mimicking the human body. The first joint is a "waist joint" that rotates the whole of the rest of the robot about a vertical axis.

Then, mounted a little to one side, is a simplified "shoulder joint." This allows the upper arm to rotate about a horizontal axis extending from the "shoulder."

Next we have a simplified "elbow joint," also allowing rotation about a horizontal axis parallel to that of the shoulder.

Then we have three wrist joints to which it is difficult to assign names. The first allows rotation about the line of the forearm, as you would use when turning a door handle. The second is a hinge perpendicular to this, such as you might use when petting a dog. The third is a twist, rather like a screwdriver held between fingers and thumb.

We need to define a chain of frames all the way from frame 0 to the gripper. For our first "journey," let us simply climb up the shaft of the robot to the height of the shoulder, where we will put frame 1. The transformation ${}^{0}_{1}T$ will convert frame 1 coordinates to frame 0 coordinates, and so will be

$$\begin{bmatrix} 1 & 0 & 0 & 0 \\ 0 & 1 & 0 & 0 \\ 0 & 0 & 1 & h \\ 0 & 0 & 0 & 1 \end{bmatrix}$$

where h is the height of the shoulder from the base. This transformation will simply add h to the z coordinate.

Now we will use the waist joint to rotate the line of the shoulder. This is a rotation about the z axis through an angle θ_1, and we will use our shorthand notation. Frame 2 will now be at shoulder height with the y axis along the line of the shoulder pivot:

$$^1_2T = \begin{bmatrix} c_1 & -s_1 & 0 & 0 \\ s_1 & c_1 & 0 & 0 \\ 0 & 0 & 1 & 0 \\ 0 & 0 & 0 & 1 \end{bmatrix}$$

Now, since the upper arm is offset from the shoulder, for frame 3 we should step in the y direction to the line of the upper arm, distance a:

$$^2_3T = \begin{bmatrix} 1 & 0 & 0 & 0 \\ 0 & 1 & 0 & a \\ 0 & 0 & 1 & 0 \\ 0 & 0 & 0 & 0 \end{bmatrix}$$

Now the shoulder axis rotates the upper arm θ_2 about the y axis of frame 3, so we align frame 4 with that limb. But should we align it with x or z? It seems logical to measure the arm's angles up and down from "straight out," so we choose x.

$$^3_4T = \begin{bmatrix} c_2 & 0 & s_2 & 0 \\ 0 & 1 & 0 & 0 \\ -s_2 & 0 & c_2 & 0 \\ 0 & 0 & 0 & 1 \end{bmatrix}$$

Now we must "move down the upper arm" to the elbow, by a distance l, say. This is in the x direction of frame 4, so

$$^4_5T = \begin{bmatrix} 1 & 0 & 0 & l \\ 0 & 1 & 0 & 0 \\ 0 & 0 & 1 & 0 \\ 0 & 0 & 0 & 0 \end{bmatrix}$$

In the Unimation Puma, the forearm is offset slightly from the upper arm, but to avoid adding an extra frame, we can take account of this in the value of a, above.

So now let us bend the elbow through θ_3 and line up frame 6 with the forearm. Once again, the pivot is the y axis and zero deflection is taken as "elbow straight":

$$ {}^{5}_{6}T = \begin{bmatrix} c_3 & 0 & s_3 & 0 \\ 0 & 1 & 0 & 0 \\ -s_3 & 0 & c_3 & 0 \\ 0 & 0 & 0 & 1 \end{bmatrix} $$

Frame 7 is lined up with the forearm, but has moved down to the wrist, distance m:

$$ {}^{6}_{7}T = \begin{bmatrix} 1 & 0 & 0 & m \\ 0 & 1 & 0 & 0 \\ 0 & 0 & 1 & 0 \\ 0 & 0 & 0 & 0 \end{bmatrix} $$

Frame 8 follows the first wrist rotation θ_4 about the local x axis.
Frame 9 "waves farewell" θ_5 about the local y axis.
Frame 10 "twists the screwdriver" θ_6 about the local x axis.
Finally, frame 11 "reaches" the tip of the "screwdriver."

As an exercise, write down the corresponding transformations.

So, just what do we do with all these matrices? Each matrix transforms the coordinates to the next-lower frame of reference; the final transformation ${}^{0}_{1}T$ brings us to the reference frame 0. But remember that the matrices are stacked up right to left, with the first to be applied closest to the vector that it multiplies, which in this case is the coordinate of a point with respect to the gripper axes. So the product ends up as

$$ {}^{0}_{11}T = {}^{0}_{1}T \; {}^{1}_{2}T \; {}^{2}_{3}T \; {}^{3}_{4}T \; {}^{4}_{5}T \; {}^{5}_{6}T \; {}^{6}_{7}T \; {}^{7}_{8}T \; {}^{8}_{9}T \; {}^{9}_{10}T \; {}^{10}_{11}T $$

We have more matrices to multiply than there are axes, but they are all elementary rotations about an axis or translation along an axis. A prismatic joint appears no different from translation along a limb. The only difference is that the distance parameter will be a variable.

Although the final matrix will be unique, there can be many ways to get there. Rotations about the y axis can be changed so that the "travel" along a limb is in the z direction, rather than x. But when the matrices are all multiplied together, they must give the same result.

There is another methodology that involves just one matrix for each actuated axis. The matrices are not primitives, as above, but are generally the product of three elementary moves.

9.3.2 D–H Parameters

The mechanism consists of a chain of links between one axis and the next. The Denavit–Hartenberg convention is based on making all rotations and prismatic actuations take place about the z axis of a frame:

- We have a set of axes at each joint. The z axes z_{n-1} and z_n at each end of link n are aligned with the axis of rotation or translation there.
- The x axis x_n at the "outer end" is chosen so that it is normal to both of these z axes.
- Now that we know x_n and z_n, we can define y_n to be perpendicular to these to make up a righthanded set of axes.
- If the z axes are not parallel, the transformation for that link must include a "twist" α about the x axis.
- The translation will consist not only of a displacement l in the x direction, but can also have a z component d to account for an offset between the points where the "previous" and the "next" normals intersect the z axis.

For a rotation θ about the first of these z axes, this results in a transformation matrix between these frames:

$$\begin{bmatrix} \cos\theta & -\sin\theta & 0 & 0 \\ \sin\theta & \cos\theta & 0 & 0 \\ 0 & 0 & 1 & 0 \\ 0 & 0 & 0 & 1 \end{bmatrix} \begin{bmatrix} 1 & 0 & 0 & l \\ 0 & 1 & 0 & 0 \\ 0 & 0 & 1 & d \\ 0 & 0 & 0 & 1 \end{bmatrix} \begin{bmatrix} 1 & 0 & 0 & 0 \\ 0 & \cos\alpha & -\sin\alpha & 0 \\ 0 & \sin\alpha & \cos\alpha & 0 \\ 0 & 0 & 0 & 1 \end{bmatrix}$$

$$= \begin{bmatrix} \cos\theta & -\sin\theta\cos\alpha & \sin\theta\sin\alpha & l\cos\theta \\ \sin\theta & \cos\theta\cos\alpha & -\cos\theta\sin\alpha & l\sin\theta \\ 0 & \sin\alpha & \cos\alpha & d \\ 0 & 0 & 0 & 1 \end{bmatrix}$$

The link transformation can thus be defined by a set of D–H parameters: the actuation angle θ, the link length l, the link offset d, and the twist α.

But with the slightest change in the convention, the "formula" for the transformation will be changed. It is my opinion that the approach of chaining a set of elementary transformations is safer and better.

9.3.3 Inverse Kinematics

Of course, calculating the kinematics of the robot is only half the story. We can now express the location and orientation of the gripper in terms of the axis movements, but what we really want is to find the axis values needed to put the gripper in some desired position. This calculation is referred to as *inverse kinematics.*

To find the required joint angles, we can calculate the transformation representing the desired position and then compare coefficients with the general transformation that is full of sines and cosines of those joint angles. That leaves us with some unpleasant simultaneous equations to solve. In fact, the result of aligning the three rotations of the wrist joint of the Unimation Puma through the same point reduces the algebra and trigonometry significantly. Nevertheless, the solutions are not unique.

For any given gripper position and attitude, there is an "elbow up" solution as well as an "elbow down" one. These are doubled again with "lefty" and "righty." By "turning its back" on the work, the robot can turn its single "right arm" into a left one.

Then, of course, not all positions have a solution. The desired position might be just out of reach of the outstretched arm.

Another problem is *singularity.* The robot normally has 6 degrees of freedom. But when two joints are in line, such as the wrist and "screwdriver twist," the degrees of freedom drop to 5. In the neighborhood of a singularity, one of the axes will have to move rapidly for the slightest change of the target position.

Think of the problem of trying to watch aircraft as they fly past straight overhead.

Of course, the robot might not have six axes, and we might not wish to move in all 6 degrees of freedom. For example, a "pick and place" robot might be concerned with placing components on a circuit board. The components are presented "flat," so we have no need to tilt them. We might, however, need to rotate them about a vertical axis to align them with the board, in which case we would need to move them to an accurate x–y position. We need a fourth axis to lift them above the board before we place them, but this might just travel between two stops.

Clearly, for a solution to make sense, there must be the same number of control axes as we wish to obtain degrees of freedom. But what if our robot has seven axes?

For various reasons, extra axes might be added, perhaps to allow the robot to "reach around corners." In this case a unique solution is impossible, not even a choice of one in four. To extract a solution, an extra condition has to be imposed, such as that one axis is held at an extreme or at zero deflection.

9.4 ROBOT DYNAMICS

From the kinematics, we have a chain of matrices that can be multiplied together to obtain the transformation matrix describing the motion of a robot. The right hand column defines the location of the origin of the gripper, while a 3×3 submatrix tells us the gripper's orientation. From this submatrix we can unravel the parameters in terms of pitch, roll, and yaw to obtain a vector with six components:

$$(x, y, z, \theta, \phi, \psi)'$$

Each of these coefficients will be a function of all six joint axes

$$x(\theta_1, \theta_2, \theta_3, \theta_4, \theta_5, \theta_6)$$
$$y(\theta_1, \theta_2, \theta_3, \theta_4, \theta_5, \theta_6)$$

and so on.

Although solving to find functions for the axis values might not be easy, we can find the effect of a "twitch" in one of the axes by partial differentiation.

If we change just θ_1 by $\delta\theta_1$, the change in x will be

$$\delta x = \frac{\partial x}{\partial \theta_1}\delta\theta_1$$

If we change more of the joints, we will have

$$\delta x = \frac{\partial x}{\partial \theta_1}\delta\theta_1 + \frac{\partial x}{\partial \theta_2}\delta\theta_2 + \frac{\partial x}{\partial \theta_3}\delta\theta_3 + \frac{\partial x}{\partial \theta_4}\delta\theta_4 + \frac{\partial x}{\partial \theta_5}\delta\theta_5 + \frac{\partial x}{\partial \theta_6}\delta\theta_6$$

In fact, we can calculate all the partial derivatives to find the Jacobian, a matrix that has these partial derivatives as its coefficients.

Now, at any given position, these coefficients will just be numbers that we can calculate, so that we can find the effect of "nudging" the joints from

$$\begin{bmatrix} \delta x \\ \delta y \\ \delta z \\ \delta\theta \\ \delta\phi \\ \delta\psi \end{bmatrix} = J \begin{bmatrix} \delta\theta_1 \\ \delta\theta_2 \\ \delta\theta_3 \\ \delta\theta_4 \\ \delta\theta_5 \\ \delta\theta_6 \end{bmatrix}$$

If we are off target, we know the values we need to approach it, at least to a first approximation. We should therefore be able to calculate a set of axis corrections to bring us closer, simply by inverting the Jacobian and multiplying by the error vector.

Often this will work! But it is possible that the Jacobian is singular and has no finite inverse. That is what happens at a singularity.

All is not lost. A method of successive approximations can bring us closer to the target, or to the point in the "reachable" space that is closest to it. For each axis in turn, inspect the corresponding column of the Jacobian and decide whether a positive or a negative nudge will bring us closer to the target, or whether that axis should remain the same. Apply the nudges and measure the new error. When there is no sign of improvement, halve the nudge size.

The Jacobian also relates the gripper velocity to the velocities of the axes. If the objective is to move it along a path at maximum speed, one or more of the axes will be required to reach maximum velocity. As the gripper moves along the path, the identity of the limiting axis will probably change. Once again, the Jacobian and its inverse will be valuable tools in calculating the axis drive values.

9.5 SIMULATING A ROBOT

Many years ago my son, Richard, helped me develop a package to simulate and articulate robot mechanisms. A version has been converted into Visual Basic and is available on the Web at http://www.essmech.com/9/5.htm.

The initial task is to design robot "parts," sets of points in three dimensions joined by a selection of lines. The data format is a set of coordinate triples defining the points and a set of integer pairs defining the pairs of points to be joined by lines.

These parts are then "assembled" to construct the robot. The robot can take the form of a simple chain, such as a manipulator, or alternatively a robot with multiple attachments such as articulated legs. The restriction is that there are no closed chains.

So, how are the parts "attached"? Two points on a component of the assembly are defined as "primary" and "secondary." They will act as the hinge about which the new part will rotate. Two points on the new part are also defined as primary and secondary where the hinge will be attached. To align the new part, the two primary points are moved together, by a simple displacement, and the new part is rotated to bring the two vectors between primary and secondary points into line. The hinge is now complete.

Each part or "limb" of the assembly now has a set of properties. First is the identifier of the shape that it takes—several parts can use the same shape. Second is the identity of the "parent" limb, the part to which it is attached, with a pair of integers to define the primary and secondary points of the parent that form the hinge. Another pair of integers will define the primary and secondary points of the part itself. Two variables describe the hinge angle and its datum value. Finally, a transformation matrix describes the absolute position and orientation of the part.

Each hinge is manipulated in turn. To change the hinge angle, a transformation is calculated that represents a rotation about the line of the hinge. This is applied to the part that "owns" the hinge and also to any other parts that are attached to that part.

The line of the hinge, defined by the primary and secondary points $\boldsymbol{p}_0$ and $\boldsymbol{p}_1$ of the parent, is found by multiplying the parent's shape coordinates by its transformation.

So, how can we find the transformation that represents rotation about this line? Let us first consider its 3×3 rotation matrix. Suppose that the direction of the hinge axis is given by the unit vector $\boldsymbol{c}$. What is the effect of rotation about this axis through the origin on a general vector $\boldsymbol{x}$?

We can break $\boldsymbol{x}$ down into a component in the direction of $\boldsymbol{c}$ and another orthogonal to it:

$$(\boldsymbol{c} \cdot \boldsymbol{x})\boldsymbol{c}$$

$$\boldsymbol{x} - (\boldsymbol{c} \cdot \boldsymbol{x})\boldsymbol{c}$$

When we rotate $\boldsymbol{x}$ about $\boldsymbol{c}$ through an angle θ, this perpendicular component will become

$$(\boldsymbol{x} - (\boldsymbol{c} \cdot \boldsymbol{x})\boldsymbol{c})\cos\theta$$

and there will be a second component the same size as

$$(\boldsymbol{x} - (\boldsymbol{c} \cdot \boldsymbol{x})\boldsymbol{c})\sin\theta$$

but in a direction perpendicular to both **c** and the orthogonal component of $\boldsymbol{x}$. But we can "turn" the orthogonal component to line up with this direction, simply by taking its cross-product with the unit vector $\boldsymbol{c}$ to get

$$\boldsymbol{c} \times (\boldsymbol{x} - (\boldsymbol{c} \cdot \boldsymbol{x})\boldsymbol{c})\sin\theta$$

But $\boldsymbol{c} \times \boldsymbol{c} = 0$, so this reduces to

$$\boldsymbol{c} \times \boldsymbol{x} \sin\theta$$

The three components can be combined to give the resulting vector

$$(\boldsymbol{c} \cdot \boldsymbol{x})\boldsymbol{c} + \{\boldsymbol{x} - (\boldsymbol{c} \cdot \boldsymbol{x})\boldsymbol{c}\}\cos\theta + \{\boldsymbol{c} \times \boldsymbol{x}\}\sin\theta$$

or

$$(\boldsymbol{c}\cdot\boldsymbol{x})\boldsymbol{c}\,(1-\cos\theta)+\boldsymbol{x}\cos\theta+\{\boldsymbol{c}\times\boldsymbol{x}\}\sin\theta$$

This is fine as a mathematical expression, but to be useful, we have to express it in matrix terms for computing. We can rearrange the first term as a matrix multiplied by $\boldsymbol{x}$. We can also express the cross-product $\boldsymbol{c}\times\boldsymbol{x}$ as the product of $\boldsymbol{x}$ with a matrix.

So, in matrix terms we have the result

$$\left(\begin{bmatrix} c_1c_1 & c_1c_2 & c_1c_3 \\ c_2c_1 & c_2c_2 & c_2c_3 \\ c_3c_1 & c_3c_2 & c_3c_3 \end{bmatrix}(1-\cos\theta)+\begin{bmatrix} 1 & 0 & 0 \\ 0 & 1 & 0 \\ 0 & 0 & 1 \end{bmatrix}\cos\theta+\begin{bmatrix} 0 & -c_3 & c_2 \\ c_3 & 0 & -c_1 \\ -c_2 & c_1 & 0 \end{bmatrix}\sin\theta\right)\begin{bmatrix} x_1 \\ x_2 \\ x_3 \end{bmatrix}$$

The braces give a "recipe" for a matrix that describes the rotary part of the transformation, R. But we need to take into account that the line $\boldsymbol{p}_0$–$\boldsymbol{p}_1$ probably does not pass through the origin, so the matrix becomes 4×4 with a translation component in the fourth column.

For the translation part, we first subtract the coordinates of the primary point $\boldsymbol{p}_0$ from $\boldsymbol{x}$, then multiply by R and add $\boldsymbol{p}_0$ again. So the fourth column of the transformation is given by

$$\boldsymbol{p}_0 - R\boldsymbol{p}_0$$

There is no need to record the bottom row of the transformation matrix, since this is always (0 0 0 1). Although needed for the perfection of a mathematician's algebra, the computer is perfectly capable of performing 3×4 matrix operations without it.

The same sort of transformation is needed to align the hinge when attaching a new part. In this case, the axis of rotation is the cross-product of the two vectors that join primary and secondary points of the component and of its parent. To calculate the angle of the rotation needed, we note that the scalar product of the two vectors divided by the product of their moduli gives us the cosine of the angle between them. The magnitude of the cross-product divided by the product of their moduli gives us the sine.

Robot joints are not always revolute. Some are prismatic, where one part slides linearly against another. This transformation is much simpler to calculate than the rotary one. It simply involves adding a proportion of vector $\boldsymbol{c}$ to every point:

$$\begin{bmatrix} 1 & 0 & 0 & kc_1 \\ 0 & 1 & 0 & kc_2 \\ 0 & 0 & 1 & kc_3 \\ 0 & 0 & 0 & 1 \end{bmatrix}$$

Now each part can be multiplied by its transformation to give its absolute position.

One way to project the coordinates for plotting on the screen is to ignore the y coordinate and simply plot the (x,z) coordinates. If desired, however, a perspective projection is simple. Plot $z/(y+r)$ against $x/(y+r)$, where r is the distance from which the robot is viewed.

Of course, the code on the Website is only the beginning. Once you have designed and tested your robot, you need to rewrite a large part of the code so that you can coordinate the simultaneous movement of the axes.

10

Further Control Theory

So far, we have followed the trend and concentrated on developing linear theory. But in the world of mechatronics, very few systems are linear. We have already seen a simulation in Section 6.5 that shows that drive limitation can totally change the way that a system performs. Nonlinearity should be a prime consideration in designing the controller. We will also find that nonlinear elements can be very useful additions to the controller itself.

10.1 CONTROL TOPOLOGY AND NONLINEAR CONTROL

10.1.1 Feedback Topology

We have examined a position control system, a second-order system with a single input. It has a characteristic equation determined by two coefficients that are set by the position and velocity terms in the equation that determined the acceleration. Putting it another way, if we decide on the roots that we want for that characteristic equation, the feedback coefficients are uniquely determined.

When there are more inputs than one, if all the state variables can be measured, we have some freedom of choice in assigning the feedback coefficients. If the system is fourth-order and has two inputs, for example, there are eight elements in the 2×4 feedback matrix. But these determine just four coefficients in the characteristic polynomial.

Essentials of Mechatronics, by John Billingsley

Some arbitrary methods can be used to give up the freedom and make a choice, but when the response is important, the matter requires careful thought, not just concerning the roots that we might want.

Consider, for example, the pitch channel of an autopilot. There are two inputs to this axis: the elevator control surface and the throttle. There are several state variables, but the pilot's concern is with the height and the airspeed. The obvious strategy is to use the throttle to control the airspeed and the "stick" to control the height. But that is not the way a human pilot thinks of it.

If the airspeed drops, with the danger of the aircraft stalling, the pilot will first push the stick forward. The opening of the throttle is a second measure that must rely on the smooth functioning of the engine for success. It makes sense to use the throttle for controlling height instead, since opening the throttle will increase the flow of energy to the system, meaning that if the aircraft maintains constant speed, it will gain height.

We have two options for the *topology* of the controller. There is the "conventional" one of feeding the airspeed back to the throttle and the height back to the elevator, or the alternative of feeding airspeed to the elevator and height to the throttle. When the system has constraints, the topology can become even more important.

Many years ago I encountered a paper-coating process. After coating, the paper passed through a drying oven before it could be cropped and stacked. A vital factor in the operation is that the paper must not stop. If it does, a large quantity of valuable product has to be scrapped and there is a risk of fire.

Somehow the flow must be maintained while a new roll is pasted on to the tail end of the previous one, and for this the system uses a "magazine," as shown in Figure 10.1.

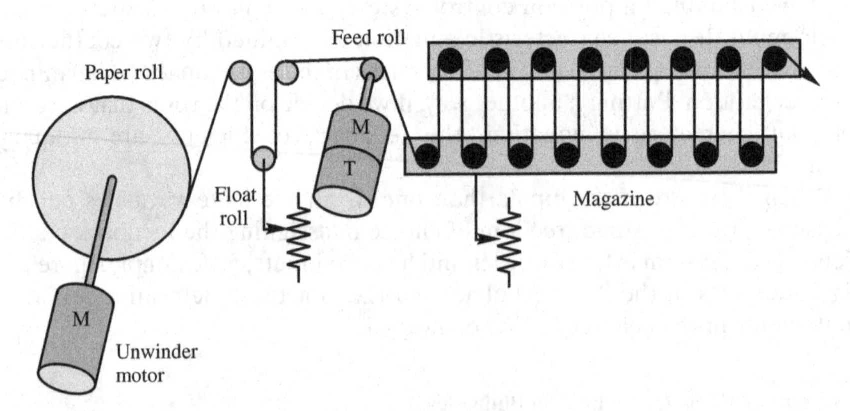

Figure 10.1 *Paper coating process.*

The magazine contains a hundred feet or more of paper. As the roll comes to an end, it is stopped and the tail clamped, while paper continues to flow out of the magazine. If all is well, the new roll is pasted before the magazine is empty and the roll can be brought up to speed. A motor with tacho feedback controls the speed at which paper is fed into the magazine, feeding 150% of the output speed when the magazine is almost empty and reducing to 90% when it is uncomfortably full.

So, where is the problem? The paper roll is driven by an "unwinder motor," which is in turn controlled by the state of a "float roll." This is a loop in the paper that takes up the fluctuation between the paper roll and the magazine feed motor. The entire variation of this loop might be two feet or less, and if it hits its stop, the whole process is halted. The restart process after pasting on a new roll must be performed with great delicacy.

The unwinder loop is indeed a difficult control problem. Although a second-order equation links motor acceleration to the float-roll position, its coefficients vary wildly. The moment of inertia of the roll will change by a factor of 60 between full and empty. Roll speed is an unreliable measure of paper velocity, so designing a system to have a good response across the range of operation is far from easy. Perhaps a change in the feedback topology can help.

As inputs, we have the drives to the magazine feed motor and to the unwinder motor; as outputs, we have sensors to tell us the positions of the float roll and the state of the magazine, plus the tacho output from the feed motor. What should we feed back to where?

The most critical item in the system is the float roll. A relatively small error can bring disaster. Which input has the most immediate effect on this roller? It is not the unwinder, to which the float roll's signal was originally applied, but the magazine feed motor. Indeed, this motor has a tacho signal that allows the float-roll control loop to be tuned to perfection. So what of the unwinder?

The tacho gives a clear measure of the magazine replenishment speed, so this can be fed back to the unwinder, mixed with the original nonlinear demand function calculated from the magazine state. If there is a large excursion in the startup transient, it is of no importance. The magazine can absorb many tens of feet of overshoot with no problem whatsoever. Figure 10.2 shows the difference between the two alternative control systems.

10.1.2 Nonlinear Feedback and Nested Loops

The "quality" of a mechatronic control system is measured not only by the way it can respond to a change in target or set point but also by the way it can withstand disturbances and recover from them.

As we saw in Section 6.5, the presence of drive limitation can completely change the rules for setting the feedback coefficients. The choice will also depend on the maximum size of the disturbance that can be expected. By introducing a nonlinearity into the feedback, the response can avoid overshoot for any size of initial error or change in demand, but if a disturbing force exceeds the full drive of the motor, it will always win the tug-of-war.

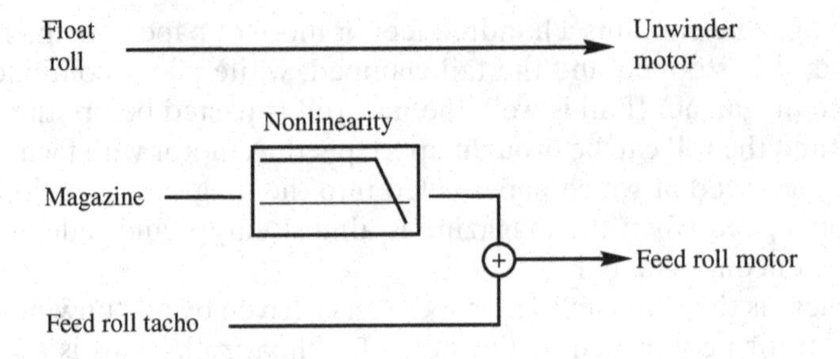

Original - with problems

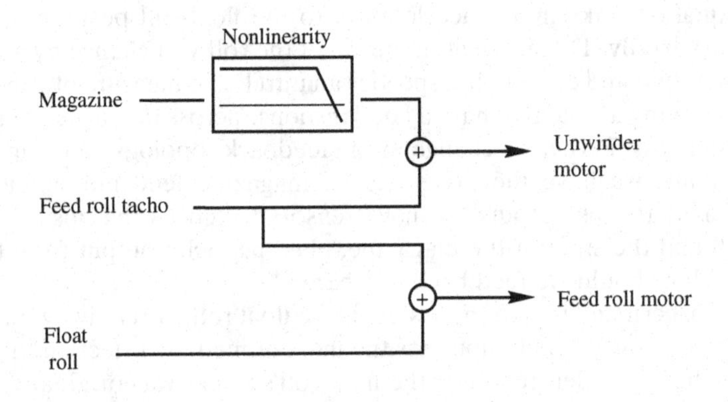

Changed topology

Figure 10.2 *Two control configurations.*

In Section 3.4.5 we experimented with a simple first-order system, the relationship between the drive to a motor and its tacho output. We were able to add a demand signal to the feedback, so that when

$$u = k(v_{\text{demand}} - v)$$

with k taking a large value, the motor will accelerate rapidly to reach the demanded value, applying full drive for any substantial error.

(In principle, the value of k can be infinite, switching the drive from one extreme to another. When the control is discrete time, however, as in computer control, a requirement for a stable response will put a limit on the feedback value that will depend on the sample rate.)

We were then able to give v_{demand} a value proportional to the position error to arrive at a closed-loop position control system. By limiting the value of

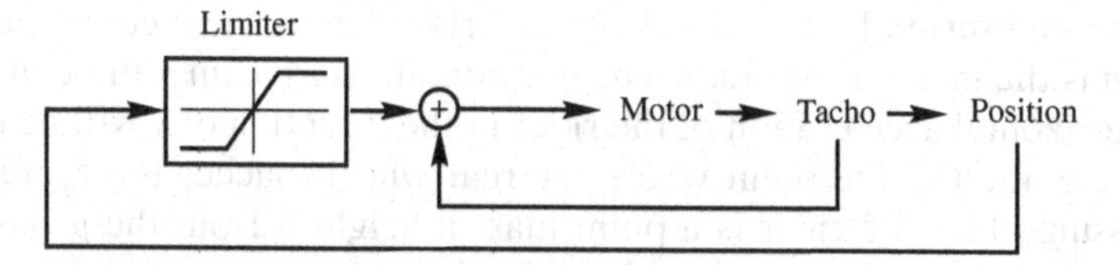

Figure 10.3 Nested loops.

v_{demand}, we were able to obtain a response without overshoot for any size of disturbance.

The control appears as two "nested loops" as shown in Figure 10.3.

In Chapter 3, we met the problem of balancing an inverted pendulum. The equations are almost identical with those that describe a bicycle that is being ridden to follow a line. Although designing a controller for a bicycle might not have great practical merit, it makes a very interesting design example.

There are four state variables that concern us. First is the distance from the line, which we will measure from the rear wheel and call x. Next is the angle of the bicycle to the line in radians, α. An important requirement is to remain upright. The angle of lean can be termed θ, and its rate of change is given the label ω. All the variables are *positive to the right.*

The input to the system is the handlebar angle u. The control task is to devise a feedback arrangement that will express u in terms of x, α, θ, and ω and give "good" control.

Let us suppose that the bicycle is proceeding at constant speed V and that each angle is small enough that its sine can be approximated to its value in radians.

First let us set up the state equations. The component of velocity perpendicular to the line is $V \sin\alpha$, so, making the approximation $\alpha = \sin\alpha$, we have

$$\dot{x} = V\alpha$$

A little geometric study will show that the rate of change of α is given in terms of V, u, and the length L between the wheels as

$$\dot{\alpha} = \frac{V}{L}u$$

When we consider the lean of the bicycle, for the first equation we have, of course

$$\dot{\theta} = \omega$$

but the second is less obvious.

The force exerted by the bicycle on the rider has vertical component mg, where m is the mass of the rider. The horizontal component will be $mg \tan\theta$, so the horizontal acceleration of the rider in the direction of x will be $g \tan\theta$. The acceleration of the point where the rear wheel touches the ground is $\ddot{x}$. If we assume that the rider is a point mass a height h from the ground, the angular acceleration is related to the difference between these two accelerations as

$$h\ddot{\theta} = g\tan\theta - \ddot{x}$$

From the first two equations involving x and α, we see that

$$\ddot{x} = V\dot{\alpha} = \frac{V^2}{L}u$$

so, making the usual approximation concerning angles, we have our fourth equation:

$$\dot{\omega} = \frac{g}{h}\theta - \frac{V^2}{Lh}u$$

In matrix form these four state equations become

$$\begin{bmatrix} \dot{x} \\ \dot{\alpha} \\ \dot{\theta} \\ \dot{\omega} \end{bmatrix} = \begin{bmatrix} 0 & V & 0 & 0 \\ 0 & 0 & 0 & 0 \\ 0 & 0 & 0 & 1 \\ 0 & 0 & g/h & 0 \end{bmatrix} \begin{bmatrix} x \\ \alpha \\ \theta \\ \omega \end{bmatrix} + \begin{bmatrix} 0 \\ V/L \\ 0_2 \\ V^2/Lh \end{bmatrix} u$$

We can substitute our feedback value for u

$$u = ax + b\alpha + c\theta + d\omega$$

to obtain the matrix for the closed-loop system. Then we can find the characteristic equation as in Section 8.1.2. It might be surprising to find that a, b, c, and d must all be positive. To move to the left, we must first turn the handlebar to the right.

The various strategies can be tried out on a simulation. To give it numerical values, set h and L both to 1 m/s and V to 2 m/s.

When we come to consider the practical implementation of a controller, we see again that the effect of constraints cannot be ignored. First there must be a limit on the handlebar angle, either from considerations of hitting the rider's knees or from the danger of skidding. The limit might be taken as 0.5 radian.

The lean angle is limited in a different way. If it is too great, the bicycle will skid and the rider will hit the ground. This is a condition that the control must seek to avoid, rather than the sort of constraint that "stabilizer wheels" would impose.

The nested-loop topology can be used to good advantage in devising a controller, assuming that we can measure or estimate all the state variables.

The first requirement is to remain upright, so this loop is closed first. But a demand signal is added into the loop as well:

$$u = c(\theta - \theta_{\text{demand}}) + d\omega$$

It is clear why c and d must be positive. If the bicycle falls to the right, the handlebars must be turned to the right. Their values can be tuned to give a rapid and well-damped correction to any disturbance, while taking the handlebar limits into consideration.

The next loop concerns the bicycle's angle to the line, α. To turn the bicycle, we require it to lean, and the lean loop takes care of the handlebars. To turn to the left, we lean to the left

$$\theta_{\text{demand}} = p(\alpha_{\text{demand}} - \alpha)$$

and to avoid disaster, we must limit θ_{demand} to, say, 0.25 radian. It is θ_{demand} that must be limited in our simulation, not θ.

If we are off the line, we demand an angle that will bring us back to it:

$$\alpha_{\text{demand}} = q(x_{\text{demand}} - x)$$

We need yet another limiter. However far we are off the line, there is a limit to the angle at which we wish to approach it.

If the system has been well designed, we can see the limits taking effect in turn as we follow a large initial offline error to the right.

First we will see the handlebars twitch to the right as the bicycle is required to lean into a turn to the left. For a short while the lean angle will be to the left at the maximum value allowed, while the handlebars are also turned to the left and the bicycle follows an arc of a circle. As the maximum heading angle is approached, the bicycle becomes upright and steers in a straight line. As the target line is approached, the bicycle leans to the right, turning to settle on the target line.

The control loop has a variable structure as each limit comes into play. Constant lean control is only a second-order system. Constant heading control is third-order and it is only when errors are small that the full four orders take effect. Provided the coefficients of θ and ω are well tuned, falling over should not be an option. A wrong choice of a and b will see the bicycle swooping from side to side, but never leaning beyond the value of the demand limit.

I first encountered the nested-loop approach when working on the design of the roll channel of an autopilot many decades ago, but the principles are still true today.

In the roll channel of a rate–rate autopilot, the tightest loop is the feedback around the aileron servomotor. The high-gain velocity control loop will cause the motor to be driven to subdue any disturbances that wind gusts might cause to the aileron control surface. A signal added into this loop will constitute a velocity demand signal.

The next loop is based on the signal from a rate gyroscope. This measures roll rate, and when fed into the aileron loop as a velocity demand, it causes the control surface to move at a rate proportional to the roll rate. The loop is closed through the response of the aircraft to the aileron control, so that a signal added into the loop becomes a roll rate demand.

Now, a passenger airliner has some serious requirements that limit the allowable maneuvers. The passengers would certainly be unhappy if the aircraft rolled at greater than 3° per second, so it is important that a demand signal injected into this loop be limited.

A position gyro that measures roll angle is the sensor for the next loop. There are a few complications associated with creating the roll rate demand from the roll angle error, but this loop now has an input that is the roll angle demand. And, of course, since to roll to an angle greater than 30° would make the passengers decidedly uncomfortable, there has to be a limiter on the demand.

When an aircraft banks (rolls), it flies in a circle. It changes its heading at a rate proportional to the roll angle, so when the pilot wishes the aircraft to fly on a compass heading, the error is fed into the roll demand.

When making the approach for a landing, a radio "localizer" beam results in a signal representing the distance off the centerline of the runway. This signal is added into the heading loop to perform the control. But when the aircraft is "acquiring" the beam, the error is large, so we wish to limit the heading change that it will cause. There is our final limiter. The resulting scheme, somewhat simplified, is shown in Figure 10.4.

On my last day with the firm before leaving for doctoral studies, I experienced a test flight with the autopilot. It worked.

10.2 PHASE PLANE METHODS

We have seen how easy it is to set up a computer simulation of a system and include constraints with a simple "IF" statement. Even so, it is useful to have non-computer-based methods for "back of an envelope" scheming.

10.2.1 Meet the Phase Plane

Many of the problems we encounter will be second-order. If we have just two state variables, such as position and velocity, we can plot the state as a point on graph. As time goes on, the state variables will change smoothly and the

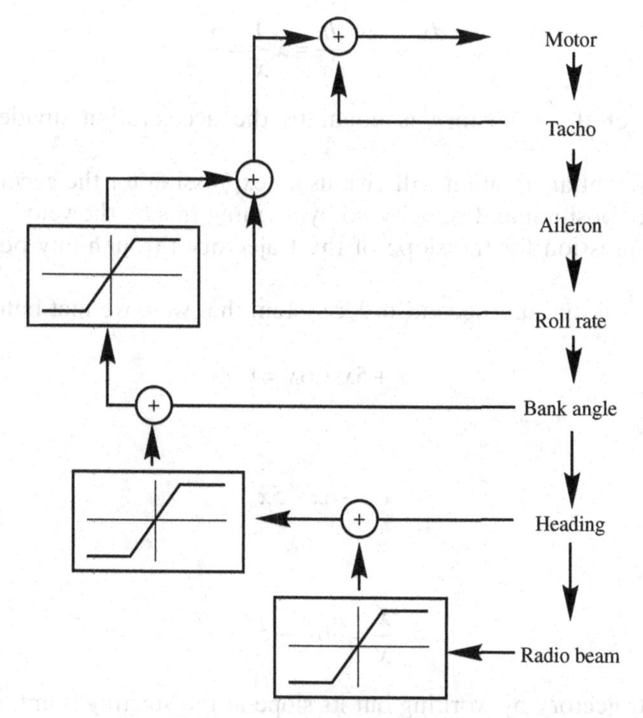

Figure 10.4 *A nested autopilot.*

point will move along a curve, a *trajectory* representing the response of the system.

If all such trajectories lead to the origin of the error–velocity plane and stay there, the system will be *asymptotically stable.* If any trajectory heads off toward infinity, we have an instability problem. But there is a third possibility. A trajectory can form a closed loop, cycling in a *limit cycle* oscillation. This can be annoying, but might not be fatal to the system meeting the design requirements. If there is a region of the plane into which all trajectories lead and from which no trajectory leaves, we have *bounded stability.*

How do we construct these trajectories? We start at some point $(x,\dot{x})$ and begin to draw the trajectory—but in which direction?

We need to know its slope. We need to know $d\dot{x}/dx$, the rate of change of the velocity with respect to the position, not to time. But maybe there is a relationship between this derivative and the time derivative.

It can be shown that if f and x are both functions of time, then

$$\frac{df}{dx} = \frac{df}{dt}\frac{dt}{dx}$$

so this is also true of $\dot{x}$, and we can write

$$\frac{d\dot{x}}{dx} = \frac{d\dot{x}}{dt}\frac{dt}{dx} = \ddot{x}\frac{1}{\dot{x}} = \frac{\ddot{x}}{\dot{x}}$$

The slope of the trajectory is equal to the acceleration divided by the velocity.

The differential equation will give us an expression for the acceleration at any value of position and velocity, so by dividing this by the velocity, we will have an expression for the slope of the trajectory through any point in the plane.

Let us try it out on a second-order system that we have met before:

$$\ddot{x} + 5\dot{x} + 6x = 0$$

Now

$$\ddot{x} = -6x - 5\dot{x}$$

so

$$\frac{\ddot{x}}{\dot{x}} = -6\frac{x}{\dot{x}} - 5$$

Tracing a trajectory by working out its slope at the starting point, drawing a small segment, working out its slope at the next point and so on threatens to be a tedious task—but fortunately there is a shortcut.

At the point (0,1) the slope will be –5. At the point (0,2) the slope will be –5. In fact, anywhere on the line $x = 0$ the slope will be –5. The line $x = 0$ is an *isocline*, a place where the slopes are all the same.

We can easily spot any number of isoclines—in this case any line on which $x/\dot{x}$ is constant, in other words, any line through the origin.

We can start straight in with the two axes. On the x axis, the slope will be infinity—the trajectories cross it at right angles. On the line $x = \dot{x}$ the slope is –11, while on $x = -\dot{x}$ the slope is 1.

We can draw these lines and mark them with small ticks in the direction of the trajectories, a sort of unfinished spider's web (see Fig. 10.5). Then we can draw the trajectory from some starting point, bending it to obey the slope as it crosses each isocline.

We need some more isoclines to get an accurate plot, especially three particular isoclines in this case.

The line

$$5\dot{x} + 6x = 0$$

is special because the acceleration there is zero. The trajectories cross it with zero slope, parallel to the x axis.

But what about the line $\dot{x} + 2x = 0$? The slope on this line is –5 – (–6/2), which gives a result of –2.

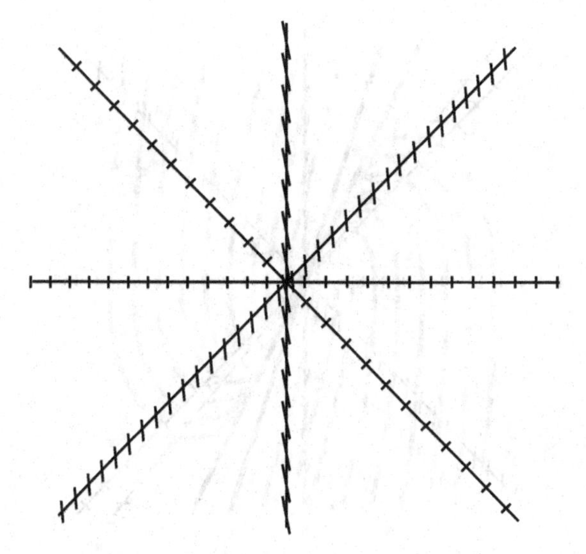

Figure 10.5 Axes and diagonals with ticks.

In other words, on the line with slope –2, the trajectories also have slope –2. Any trajectory reaching or starting on this line will be "glued" to it.

We find that the same is true for the line

$$\dot{x} + 3x = 0$$

After drawing these onto the diagram, a good sketch of the phase plane can be created (see Fig. 10.6).

Let us take another look at these "special" isoclines. If the trajectory follows

$$\dot{x} + 2x = 0$$

this is not just the equation of a line; it is a differential equation. It tells us that

$$x = x(0)\mathrm{e}^{-2t}$$

which should not be surprising, since we have already found the general solution of this differential equation to be

$$x = A\mathrm{e}^{-2t} + B\,\mathrm{e}^{-3t}$$

The two special isoclines are the special cases where A or B is zero. But we can learn a little more. As time advances, the e^{-3t} term will decay faster than the other term, so the trajectories will become asymptotic to the line

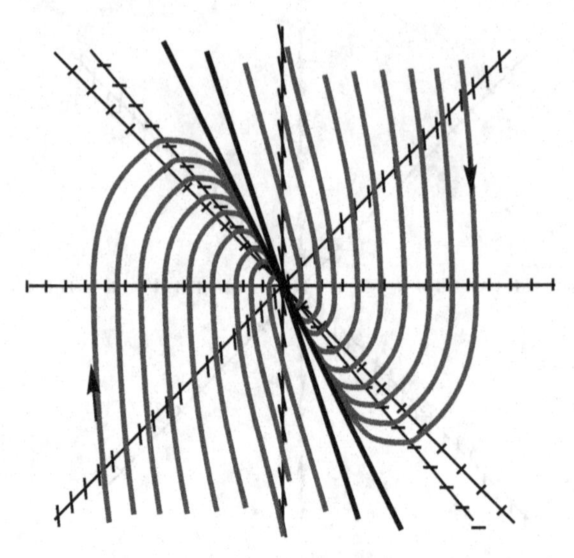

Figure 10.6 Phase plane sketch.

$$\dot{x} + 2x = 0$$

On the other hand, if we trace the trajectories backward into the past, the e^{-3t} term will become dominant, so the slopes will become asymptotic to –3.

As an exercise, sketch the phase plane for the system

$$\ddot{x} + 3\dot{x} + 6x = 0$$

When you solve the characteristic equation to find the "special" isoclines that are asymptotes, you will find that the roots are complex. These isoclines do not exist. The system is underdamped and the trajectories perform spirals around the origin as the system "rings."

10.2.2 Dealing with Constraints

We seem to have devoted considerable effort to deal with a system that we had already analyzed analytically. But the analytic method is tailored only for linear systems. The phase plane comes into its own when there are constraints and other nonlinearities.

Let us again consider the system

$$\ddot{x} = u$$

where

$$u = -5\dot{x} - 6x$$

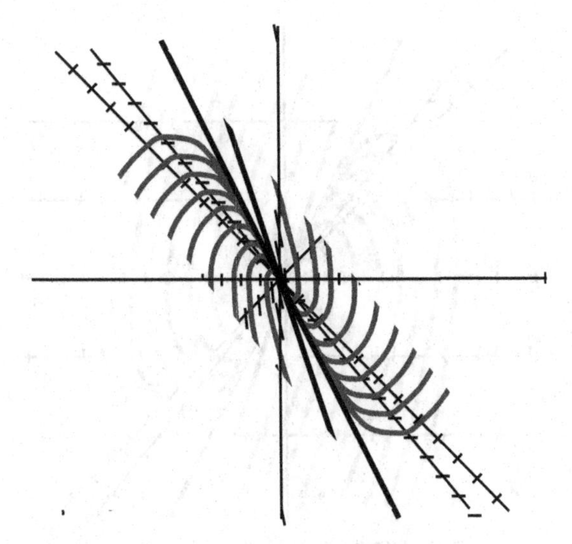

Figure 10.7 Linear region of phase plane.

But this time we have a limit on u, namely, $|u| \le 4$.

Close to the origin the drive is not saturated and the phase plane is just as we have drawn it, but outside the linear region the system equation is

$$\ddot{x} = 4$$

or

$$\ddot{x} = -4$$

The boundary of the linear region will be the two lines

$$-5\dot{x} - 6x = \pm 4$$

These are parallel to the line on which the drive is zero, so the linear region appears as shown in Figure 10.7.

In the two other regions, the slope is given by

$$\frac{d\dot{x}}{dx} = \frac{4}{\dot{x}}$$

and

$$\frac{d\dot{x}}{dx} = -\frac{4}{\dot{x}}$$

The isoclines in both cases will be lines of constant $\dot{x}$, parallel to the x axis.

The trajectories will be parabolas, and the complete phase plane will resemble Figure 10.8.

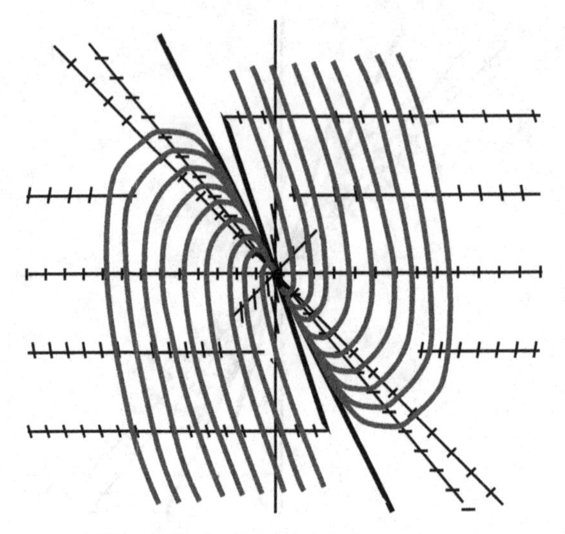

Figure 10.8 *Composite phase plane.*

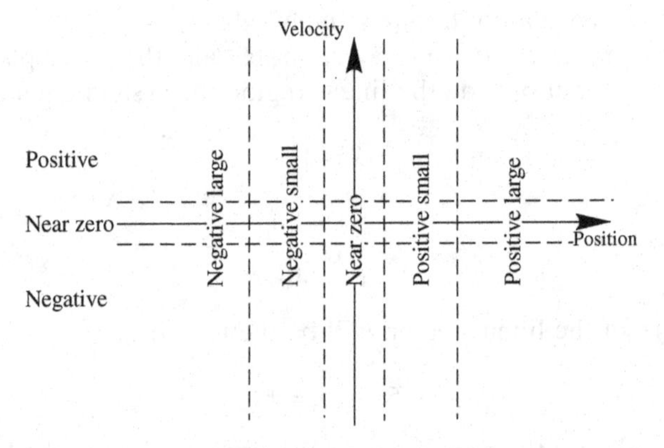

Figure 10.9 *Fuzzy logic phase plane.*

The phase plane can deal with a wealth of nonlinearities, including friction and deadband and, with a little ingenuity, backlash. It can also be used to try out a variety of nonlinear feedback strategies.

One approach that has been fashionable is termed "fuzzy logic." Instead of a precise measurement of position and velocity, their values are simply reported in terms such as "near zero," "positive small," "positive large," and so on. These divisions will divide the phase plane into a tartan pattern of combinations of position and velocity ranges similar to that shown in Figure 10.9. In each rectangle, the control designer can put any available value of drive. On one hand this will overcome the weakness of linear feedback strategies, but on the other hand there are some serious limitations on performance.

The system can come to rest anywhere in the near-zero range of x with no further attempt at correction, leaving a standing error. To avoid this, the near-zero range can be omitted, leaving the zone boundary as the axis. But now we can be left with a limit cycle or at best multiple overshoots as the drive switches to and fro.

If the velocity is added to the position as continuous signals before the values are cropped into ranges, we can, of course, have a crisp nonovershooting response. But this is not how the game is usually played.

10.3 OPTIMIZATION

If the requirement is simply one of stability, there are untold possible variations in the feedback parameters. The "by the book" control system designer would like to find a unique solution that is somehow "the best." This is optimal control.

10.3.1 Least Squares

If we are seeking the best, we must have a measure of the quality that we are trying to optimize. The problem is described in the form of a *cost function*, and the design task becomes one of minimizing that cost function. The cost function could be something explicit, such as settling time or fuel consumption, but a textbook favorite is *least squares.*

For the second-order system

$$\ddot{x} = u$$

we might choose a cost function

$$C = ax^2 + b\dot{x}^2 + cu^2$$

After some mathematical manipulation, we discover that the controller that minimizes the integral of C is based on proportional feedback. Indeed, if $b = 0$, so that the cost involves only the position and the input, the solution has a damping factor of 0.707.

Examples can be found in process control, where "gentle" adjustment is in order. The controller acts as a "regulator," keeping the process at an optimal setpoint while countering any disturbances.

However, we have already seen that if the input is limited, as in a servomotor, the design should depend heavily on that limit. Selecting a quadratic cost function as the basis from which to design the controller loses any logical reason when the coefficients of the cost function have to be "fiddled" to give an acceptable response.

10.3.2 Time-Optimal Control

When we have a "real" cost function, such as the time to reach zero error with all derivatives at zero, the solution is usually *bang-bang* control. The input is at all times at one or the other limit until the target is reached. Techniques such as *dynamic programming* and the *maximum principle* can define the nature of the switching function, but it takes more ingenuity to find the actual switching times that will bring all the errors simultaneously to zero.

For our simple example of acceleration control, the *maximum principle* deduces that just one change of sign is required to bring the system to the target. But when? Imagine that you are driving from one traffic light to the next in minimum time, in a vehicle with just one gear and insufficient power to "burn rubber." As the light turns green, you must put the pedal to the floor. At some point before hitting the next red light, you must apply full brakes. When?

For minimum time, you must hit the brakes at the last possible moment from which you can actually stop before the light. Your *time-optimal trajectory* consists of a period of maximum acceleration switching to a period of maximum deceleration, coming to rest at the target when zero drive is applied. The "quality" of your control depends on your ability to model the braking process accurately with some sort of switching curve. If your estimate of your braking power is overoptimistic, or if there is any disturbance that pushes the car on its way, then an overshoot is unavoidable.

If, on the other hand, the braking deceleration is underestimated, the settling time will be slightly increased but there will be leeway to account for mishaps. The nature of the control will be *sliding*. Brakes will be applied as the switching line is crossed, but the greater-than-predicted deceleration will take the state across the line again and acceleration will be applied. The drive will switch rapidly to and fro, causing the state to follow the switching line. There is a simulation example at www.essmech.com/10/3/2.htm.

Time optimization has much in common with the task of fuel optimization in a lunar lander. Many years ago, our Cambridge group received a visitor from Moscow. He told of the computational task of calculating the control to bring the first unmanned lunar probe to rest with minimum fuel consumption.

The nature of the solution was the same as that for time-optimal control. The probe is allowed to fall freely until the last moment, when continuous full drive will just bring it to rest as it touches the surface. Unfortunately, at that time the first two or three probes had landed far from softly.

At the time that the final burn is initiated, the probe might be falling at a mile per second. A one-second error will leave the probe irrevocably heading on a trajectory that would end a mile beneath the surface, if an impact did not intervene!

My suggestion, that a deliberate underestimate of the thrust would cause a minimal increase in the fuel actually consumed, was passed on very tactfully by our professor. The next probe landed successfully.

Optimization might serve the purpose of giving a unique solution that can be claimed to be "right," but it is seldom the best in practice. The function that needed to be optimized in this particular case was the probability of a "successful" landing. Any fuel that remained after the landing was of no value whatsoever.

11

Computer Implementation

Having devised control algorithms and converted them to software of one form or another, our next step is to integrate the system to include a computer to run it on. For the experimental work, it has been easiest to exploit a retired PC, but for serious product development, some sort of computing engine must be integrated into the design as a whole.

There are some features that are common to the PC and to the humblest of microcontrollers, which can greatly influence your approach to the task.

11.1 ESSENTIALS OF COMPUTING

As the computer has evolved, many ingenious variations have been tried. Some have survived, while some have gone the way of the dodo. But some underlying principles remain unchanged.

11.1.1 General Fundamentals

The simplest computing engine is the *Turing machine*. This is really a figment of the mathematicians' imagination, used to decide what is "computable" or not. It has an input bit and a "state" signifying which "instruction card" is in play. From these, the output bit and the next instruction to be used are specified.

When we get to a "real" computer, the essentials are memory, program counter, and, of course, input and output. There are usually several sorts of memory. The most easily accessible are "registers" such as one or more accumulators to hold the value that any calculation has reached so far and RAM (random access memory), an array of "pigeonholes" in which numbers can be stored.

Any embedded system also has ROM (read-only memory) that contains code and data that cannot change. To complicate matters, there is also EAROM (electrically alterable ROM) to hold data that must survive the system being switched off.

We now have a program that is stored in memory. In Von Neumann machines, the great majority, this memory can double for both program and data, although some other devices have separate memory formats. To access a byte or word of memory, its *address* is placed on a *memory address bus.*

Most instructions will manipulate data, performing arithmetic or logic operations on values in the memory or registers and going on to execute the next instruction. The address of this instruction is held in the *program counter.* Other instructions will influence the program flow, with *branch* or *jump* instructions to allow a piece of code to be skipped or executed repeatedly. There are also *conditional jumps* to determine whether to jump according to the result of a comparison, so that a loop can be terminated after a number of executions or on the result of some input value.

Input–output is, of course, a vital operation without which the computer has no real purpose. The "classic" form of input is to transfer 8 or 16 bits, represented by logic voltages on an array of input connections, to an accumulator register within the processor. Input–output registers are often *memory-mapped*; in other words, they behave as though they are memory at some specific location, enabling values to be input and output as though reading from or writing to memory.

11.1.2 Subroutines

In the evolution of the computer, an important "bright idea" was the conditional jump. Another was the subroutine call. Suppose that we wish to perform a special operation on a number, such as evaluating its logarithm. We can write a block of code to perform the operation and include it in the software. Now suppose that we wish to evaluate the logarithm of another number, somewhere else in the program. We could, of course, plant a second copy of the logarithm code in the program, but this would waste space.

Instead we can "call" a single copy of the logarithm routine from a number of different parts of the program. This is different from a "jump," since we must know where to return afterward. We must also tell the routine the value that we wish to convert, and in turn the routine has to convey the answer when returning. This could be done by holding values in the registers, but the accepted method is to use the "stack."

Most processors have a *stack pointer*, holding a value that points to an area of memory which is otherwise uncommitted. A PUSH command will save a register's value in the address pointed to by the stack pointer, which will automatically increment (or decrement) to point to the next location. A POP command will do the opposite, reading from the stack (after decrementing or incrementing) and restoring the register. Various processors work in various ways, incrementing or decrementing before or afterward, but the programmer simply has to ensure that the PUSHes match the POPs.

Now, in a subroutine call, the program counter is pushed onto the stack, and its value is retrieved when the code "returns."

11.1.3 Interrupts

The next bright idea was the interrupt. Until then, the program execution had depended on the program itself, together with any values that are input and later used to influence conditional jumps. At various places in the software, a call might be made to a subroutine to check whether a new byte was ready for input. A loop that checks inputs in turn is termed a *polling loop*.

With an interrupt, the data-ready event grabs the attention of the computer and takes it immediately to the routine that will deal with it. The machinery pushes the return address onto the stack, together with the status register. The interrupt routine then starts to execute. Its first task is to save any register that might be changed in the routine, so that afterward the computer can pick up the action where it left off, just as though the interruption had not happened.

The interrupt can be caused by the arrival of data, by an external device being ready for another byte of output, by some sort of timer, or by an input event such as the pressing of an emergency button.

Despite their great advantages, interrupts are a nightmare for real-time troubleshooting. Except under the most artificial of conditions, the program will never be executed the same way twice. What is more, interrupts lead to a multitude of philosophical problems for the operating system designer. What happens if the computer is executing one device's interrupt routine when another interrupt arrives?

This leads to the idea of an interrupt hierarchy. An interrupt is permitted only if it is "more important" than any interrupt state existing.

But the philosophy gets deeper. When a data byte arrives on a high-speed serial connection, there is an urgent need to read it before the next byte arrives. The interrupt routine's purpose is clear. It must copy the byte to a "buffer" in memory, and then computing can resume.

What happens when the last byte is received and the data transfer is complete, however? How does the software decide which task should take precedence?

But first there is a faster means of dealing with data transfer: direct memory access (DMA).

11.1.4 Direct Memory Access

Some devices are capable of sending a burst of data. It is a wasteful operation to execute an interrupt for each byte, with the need to save registers and "environmental variables," then to input the byte and save it in the correct location, and then to restore the variables and return. Instead, DMA allows an external device to gain access to the memory address lines and plant the data in memory with no reference to the processor itself.

A "bus request" is pulled down, and when a "bus grant" is given, the transfer can begin. In the case of the PC, the term "external" is a relative matter. A DMA controller is built into the hardware and performs all the hard work. This is primed with the start address of the memory to be filled and the number of bytes or words to transfer. It clocks new bytes from the peripheral and saves them sequentially in memory. When the last has been received, it releases the bus request and, if desired, causes a hardware interrupt so that the data can be dealt with.

Of course, the process can work in reverse with the contents of a memory block being output.

11.2 SOFTWARE IMPLICATIONS

The ways of programmers are something of an enigma. On one hand, the GOTO statement is deprecated, for very sound reasons, while on the other hand flow diagrams are encouraged—yet every line in the flowchart is the embodiment of a GOTO statement!

In the early days of programming, overenthusiastic software writers were often guilty of "spaghetti code," with jumps in and out of loops that required great patience to trace. Another vice was the use of identifiers such as `a` or `i5`, which gave no clue as to their purpose or meaning.

But surely the pendulum has swung too far the other way, when identifiers such as

```
CoGetInterfaceAndReleaseStream
CoMarshalInterThreadInterfaceInStream
StgGetIFillLockBytesOnILockBytes
CoGetCurrentLogicalThreadId
WdtpInterfacePointer_UserMarshal
WdtpInterfacePointer_UserUnmarshal
```

are quite typical within a popular operating system.

Software tools allow great slabs of code to be stacked up in a pile that defies the efforts of the programmer to read through and check the fundamental details. But when real-time code is to be written for an embedded machine, there are great virtues in keeping it lean and mean. It is my opinion that wherever possible, identifiers should be no longer than two syllables, whether they are variable names or procedures.

The choice of a computer language is not a simple matter. Language has many dimensions. First there is the "speak," the words and symbols that will be used to define the code. A page of Java will look very much like a page of C, while the line-by-line text of a Visual Basic program will look very much like other forms of Basic—and might even bring back distant memories of FORTRAN.

Underlying the code is the structure of its execution. What makes Visual Basic visual is its use of "forms" on which are placed "controls." Each control has a piece of code to deal with any "event," such as the click of a mouse, the operation of a "button," or the change in a "slider." This has every appearance of being real time, but in most cases the interrupts are illusory. There is an instruction, `DoEvents`, which really means "go and poll any other tasks that might need attention." A loop that does not include a `DoEvents` can lock up the machine so that user inputs are ignored.

QBasic or Quick Basic, on the other hand, will allow an event to grab control at the completion of any instruction. In general, the "lower" the level of a language, the more control the programmer will have over the way the code will be executed. C is close to assembly language and allows much greater control.

The "programming environment" is another important factor. A "user-friendly" system will check each line as it is entered and signal any syntax errors. In Visual Basic, a click on the RUN icon is all that is needed to test the code. When the program crashes, moving the cursor to any variable will cause its value to be shown in a TOOL TIP, while the offending line of code is highlighted in yellow. The programmer is faced only with the task of entering code that is "obviously necessary," plus the properties of controls, such as their background color, and the layout of the forms.

At present there are at least two "flavors" of C++ language. The Borland version leans toward Visual Basic, with controls that can be dropped into forms resulting in the automatic generation of the associated code. In the Microsoft version, there is much more housekeeping to do. First you must decide on what sort of project you require. Is it "bare screen," or do you wish to have forms and controls? Will it result in an exe file, a dll library, or a DirectX filter?

Then, after a "wizard" has set up the empty project files for you—although they might already look pretty crowded—you have to be concerned with both code files and "header" files that define how your functions are to be called.

Before you can run your project, you have to "build" it. Only then do you see a list of your errors. The omission of a single } brace can result in a list of a dozen or more errors. The "friendly features" are not made very clear in the documentation or help files, but after much exasperation you find that clicking on a line warning you of an error will actually take you to the offending line itself. By inserting breakpoints, you can achieve the same display of variable values that Visual Basic offered so easily.

So, why endure the hardships of riding bareback? C gives access to the "inner workings" of the machine in a way that is protected from VB users. Its closeness to machine code allows it to perform tasks that in VB would require the writing of special library routines—and these would probably be written in C in preference to assembler.

In C, you certainly have more control of the way the code is executed, but the promise of more efficient code than in Basic might be a false one. Consider the artificial and useless piece of Basic code:

```
DEFINT A-Z
DIM a(10),i
i = 5
a(i) = a(i) + i
```

The array and a variable are defined as integers, the value 5 is placed in *i*, and then the *i*th element of *a*( ) has *i* added to it. It looks as though the pointer into the array will have to be calculated twice and that a C version could be much more efficient:

```
void main(){
int a[10], i;
i=5;
a[i]+=i;
}
```

The cryptic final line of the C version certainly looks more compact. But Quick Basic has an optimizing compiler. When the resulting assembler code is listed, we see

```
mov  I%,0005h
mov  si,I%
sal  si,1
mov  ax,I%
add  A%[si],ax
```

which really could not be more efficient. In the last line, the value of *i* is added straight into the array, using an index that has been loaded with the correct value, then shifted left to make it a word pointer.

The C version is converted by Visual C++ to give

```
mov   WORD PTR _i$[ebp], 5
movsx eax, WORD PTR _i$[ebp]
mov   cx, WORD PTR _a$[ebp+eax*2]
```

```
add    cx, WORD PTR _i$[ebp]
movsx edx, WORD PTR _i$[ebp]
mov    WORD PTR _a$[ebp+edx*2], cx
```

The addition is performed in a register that then has to be saved.

Writing code for a simple embedded processor is likely to have even fewer home comforts. It will involve first keying in the code as a text document. The assembler (or maybe a C compiler) is then invoked to convert the code and produce a binary "object file." This must then be downloaded to the processor in yet another operation. This must all be achieved before the code can be tested, and additional means must be devised for monitoring what the software is actually doing.

11.2.1 Structured Code

We have seen that a subroutine or "procedure" economizes on the space of code that might otherwise have to be repeated. It has another important role, however. It is a module of code that can be tested exhaustively and can then be called with a simple well-named command. A lengthy matrix inversion routine could appear in the program flow as just

```
invert a()
```

When the code is written at the assembler or C level, good structure is even more important. It is also essential to document the code with clear comments. The choice of identifiers can do much to improve readability, using verbs for procedures and nouns for variables.

I am greatly in favor of the use of "pseudocode," a language that exists in the mind of the program writer. Consider the task of writing code for a four-legged walking robot with vacuum grippers on its feet. A pace could involve moving each foot forward in turn and might be represented as follows:

```
Sub Pace()
   For foot = frontleft to hindright
      Lift foot
      Advance foot
      Place foot
   Next foot
End Sub
```

At this level, before getting bogged down in software details, it will be clear that taking a second pace will present the problem that all four feet are already in the forward position!

So it is no effort to make a note that when writing the Advance subroutine, while "this" foot is being moved forwards the other feet must be moved one-third of a stride to the rear.

In fact, this code example could become Visual Basic code as it stands. For other implementation, such as in assembly language or C, it could appear as the "remarks" that are added to make the code meaningful.

```
*      Sub Pace()
PACE   LDAA #FRONTLEFT  * For foot = frontleft
       STAA FOOT        *    to hindright
PACE1  JSR  LIFT        * Lift foot
       JSR  ADVANCE     * Advance foot
       JSR  PLACE       * Place foot
       INC  FOOT        * Next foot
       LDAA FOOT
       CMPA #HINDRIGHT
       BLE  PACE1
       RTS              * End Sub
```

In this example, the pseudocode can burrow down a level to define

```
SUB lift(foot AS INTEGER)
   Unstick foot
   target(foot).z = target(foot).z+100
END SUB
```

in which we assume that there is an interrupt routine running in the background that handles position control.

The `Unstick` routine will output the signal that releases the vacuum, then will pause for an instant. The `target` line could instead output a value to an independent microcontroller dedicated to the control of that particular leg.

11.3 EMBEDDED PROCESSORS

While the microprocessors at the heart of a personal computer are still evolving in speed and complexity at an increasing rate, some of their humble cousins have remained in fashion for much longer. These are the simple processors that are embedded in washing machines, toys, clock radios, automobiles, and a host of other appliances.

A general electronics catalog has over 50 pages of microprocessors and microcontrollers, some of them from families virtually unchanged over 20 years. Of course, there have been numerous innovations, such as the ability to communicate over USB and CAN-bus, but assembly code written years ago can often still be adapted with relative ease.

The PC user may grumble at the length of time taken for the system to load, but will not consider the "boot" process as a personal worry. For the designer of an embedded system, the entire startup process from the first application of power must be part of the design.

11.3.1 Essentials of a Microprocessor

The early 8-bit chips consisted of the processor alone, plus a number of internal registers. Random access memory (RAM) for calculations and read-only memory (ROM) to hold the program had to be wired on to address and data buses, usually with extra chips for address decoding plus a crystal to set the clock frequency.

Many of today's chips aim for an "all in one package" approach. They may have 256 bytes or more of RAM, sufficient for many embedded tasks, plus several kilobytes of program memory. For laboratory and development work, this can conveniently be EAROM, electrically alterable ROM. The contents of the EAROM can be changed by the processor itself, although this is often much slower than normal RAM operations. The data will remain unchanged when the processor is switched off and on again.

This is a comfortable size of program to handle in assembly language, but much larger memories are common. A "thumb drive" or MP3 player with less than 64Mbyte of memory would be regarded as tiny.

Just as programming languages have their faithful adherants, enthusiasts will concentrate on a particular microcomputer system. My colleague, Mark Phythian, is especially supportive of the PIC computer. He has designed the simple application described below, in which it serves as an analog-to-digital converter, encoding and transmitting the results over a serial interface to the host PC.

By using this interface, many of the problems of the Windows operating system are bypassed. A Visual Basic program is outlined below that includes an MSComms component to handle the serial communications. The values of four channels are displayed on a form as a sort of "oscilloscope trace". From here it is a small step to using the chip for online control.

The chip is a PIC16F88 from Microchip. It has four 10-bit analog input channels, 4 bits of logic input, and 4 output bits. It also has two interrupt lines and a bidirectional serial interface. Two transistors with four resistors and a diode are needed to convert the serial logic levels to something compatible with the PC's RS232 interface, but with those the circuit is complete.

The circuit diagram in Figure 11.1 shows a potentiometer and pushbuttons that can be used to test the circuit's operation, but these are not part of the basic design.

When writing code from first principles, you can make up the command rules as you go along. But it is important that they be strictly structured in a well-defined protocol.

This particular software operates as follows. The PC sends a command as a single byte. The PIC responds with the same byte, followed by any data bytes that are requested.

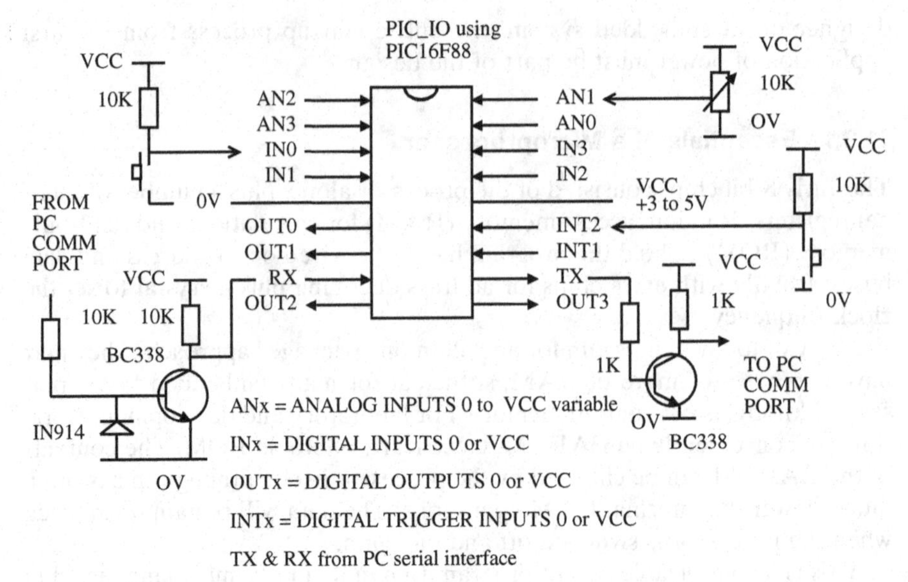

Figure 11.1 *Mark Phythian's circuit of single-chip microcomputer ADC.*

Hex \$30 returns 8 bytes, representing four channels of 10-bit ADC readings.

Hex \$40 returns 4 bytes, representing four channels of 8-bit ADC readings.

Hex \$50 returns 1 byte, with the lower 4 bits representing the input bits.

Hex \$60 to \$6F will cause the lower 4 bits to be sent to the output pins.

For the two interrupt pins on the PIC (active low):

INT1 sends 1 byte hex \$21 character.

INT2 sends 1 byte hex \$22 character.

Even with such a simple chip, it is not necessary to revert to assembly language. A Basic *cross-compiler* can be purchased for about thirty Australian dollars from http://www.oshonsoft.com/, and it is for this system that Mark has written his code.

Even so, it is necessary to attend to every detail of setting up the chip's state, defining variables and enabling the necessary interrupts to handle communications:

```
' PIC_IO_BAS.bas PIC serial IO interface by Mark Phythian
' Uses Microchip PIC16F88 processor running at 8MHz
internal RC Osc.

' define variables
```

```
Dim adtable(4) As Word ' table to hold adc results
Dim chan As Byte       ' channel number
Dim val As Word        ' word size adc result variable
Dim oldb As Byte       ' last portb value
Dim char As Byte       ' single character command
Dim command As Byte    ' upper 4 bits of command
Dim n As Byte          ' lower 4 bits of command
Dim m As Byte          ' byte size temporary variable
Dim wrd As Word        ' word size temporary variable

' setup PIC
OSCCON = 0x72          ' set internal RC select to 8MHz
TRISB = %11000000      ' set PORTB 0-5 pins as outputs,
                       '6 & 7 inputs
Gosub initad           ' initialise adc
PIE1.RCIE = 1          ' enable UART RX interrupt
OPTION_REG = 0x7f      ' enable PORTB weak pullups for
                       inputs 6 & 7
Hseropen 57600         ' set UART BAUD rate
INTCON = 0xc8          ' enable GIE, PEIE and RBIE

' initialise variables
PORTB = 0x00
oldb = PORTB And 0xc0  ' last value of PORTB bits 6 &
                       '7 for change of state
WaitMs 1000
Hserout "OK"           ' startup ok
WaitMs 100

' endless loop converting as fast as possible
main:

ADCON0 = 0xc1            ' select channel 0 in bits
                         '5,4,3 of ADCON0

For chan = 0 To 3
  Gosub adconv         ' go to conversion routine
  val.HB = ADRESH      ' save high byte (upper 2 bits
                       'only)
  val.LB = ADRESL      ' save low byte
  adtable(chan) = val
  ADCON0 = ADCON0 + 0x08 ' increment selected channel
Next chan
Goto main 'repeat forever
End
```

```
' Initialise ADC
initad:
TRISA = %11111111          ' set portA as input
ANSEL = %00001111          ' set PORTA pins 0-3 as analog
                           'inputs
ADCON1 = 0x80              ' set 10 bit A/D result format
                           ' right justify ADRESH/L
ADCON0 = 0xc1              ' set A/D conversion clock to
                           'internal source,
                           ' turn on adc
Return

' Adc conversion routine
adconv:
High ADCON0.GO_DONE        ' start the conversion
While ADCON0.GO_DONE       ' wait until conversion is
                           'completed
Wend
Return

On Interrupt
Save System
' check for PORTB change of state on bits 6 & 7
If INTCON.RBIF = 1 Then ' test if portb change flag is on
  n = PORTB And 0xc0
  INTCON.RBIF = 0          ' reset RBI flag
  n = oldb Xor n
  If n.7 = 1 Then ' if bit 7 changed
    If oldb.7 = 1 Then ' if bit 7 changed to 0
      Hserout 0x22         ' send a " for INT2 input trigger
    Endif
  Else
    If n.6 = 1 Then        ' if bit 6 changed
      If oldb.6 = 1 Then ' if bit 6 7 changed to 0
      Hserout 0x21    ' send a ! for INT1 input trigger
      Endif
    Endif
  Endif
  oldb = PORTB And 0xc0 ' set oldb to new PORTB value
Else

' test for serial command received
  If PIR1.RCIF = 1 Then ' test if RXer flag is on
    PIR1.RCIF = 0          ' reset RCI flag
    Hserget char           ' get the received character
```

```
    command = char And 0xf0 ' command is upper 4 bits
    n = char And 0x0f   ' number is lower 4 bits

' fetch 10 bit adc values, returns 2 bytes each,
' command letter 0 (zero)
     If command = 0x30 Then
       Hserout char     ' echo command
       For n = 0 To 3
          Hserout adtable(n) ' send 10 bit value in 2
                              'bytes
       Next n
     Else

' fetch 8 bit adc values, returns 1 byte each, command
letter @
       If command = 0x40 Then
         Hserout char         ' echo command
         For n = 0 To 3       ' D command requests all 4
           wrd = adtable(n)
           m = ShiftRight(wrd, 2)
           Hserout m          ' send 8 bit value as 1 byte
         Next n
       Else

' read inputs bits PA4-7, returns 1 byte, command
'letter P
          If command = 0x50 Then
            Hserout char ' echo command
            n = PORTA And 0xf0
            n = ShiftRight(n, 4)
            Hserout n          ' send 4 bits in low part
                               'of byte
          Else

' set outputs, command letters (from $60-$6F)
',a,b,c,d . . . o
            If command = 0x60 Then
              Hserout char ' echo command
              m = n And 0x03 ' arrange bits to Port B
              bits 4,3,1,0
              n = n And 0x0c
              n = ShiftLeft(n, 1)
              PORTB = m Or n ' set output bits from
              number n
            Endif
```

```
        Endif
      Endif
    Endif
  Endif
Endif
exit:
Resume
```

This code and the code for the Visual Basic test program can be found at www.essmech.com/11/3/1.htm.

The Visual Basic form has an MSComms control named `Serial` and a button with the name and caption `Quit`. Its code is as follows:

```
Dim bits10 As Byte          'For holding command
                            definitions
Dim bits8 As Byte
Dim getpins As Byte
Dim setpins As Byte
Dim bytes_in() As Byte
Dim Adc(3) As Single        'To hold ADC values between
                            '-1 and 1
Dim stopped As Boolean

Private Sub Form_Load()     'Execution starts here
Dim i As Integer
Show
Serial.Settings = "57600,n,8,1"          'make sure same
                                         'baud as PIC
Serial.CommPort = 1
Serial.Handshaking = comNone             'no handshake
Serial.InputMode = comInputModeBinary    'not ASCII text
Serial.NullDiscard = False               'treat nulls as
                                         'valid characters
Serial.PortOpen = True                   'open port

bits10 = &H30   'encode 4 channels, return 8 bytes of
                '10-bit data
bits8 = &H40    'encode 4 channels, return 4 bytes of
                '8-bit data
getpins = &H50  'read input pins, return in lower four
                'bits
setpins = &H60  'add required bit values to the lower 4
                'bits

Print "Cannot find PIC"    'Write warning message
Command setpins            'This will hang if PIC is not
                           present
```

```
Scale (0, 1)-(1000, -1)
Cls                         'Erase the message if all OK
stopped = False
Do Until stopped
   For i = 1 To 1000
      Adc10                        'contains DoEvents
      PSet (i, Adc(0)), vbBlack 'Plot the ADC values
      PSet (i, Adc(1)), vbRed
      PSet (i, Adc(2)), vbBlue
      PSet (i, Adc(3)), vbGreen
   Next
   Cls                             'Clear at end of trace
Loop
End                                'End if the loop exits
End Sub

Sub Command(a As Byte) 'will flush buffer if necessary,
                       'hang if no PIC
Dim b(0) As Byte
b(0) = a
send b()
Do                                 'Wait for the echo byte
   get_bytes 1
Loop Until bytes_in(0) = a
End Sub

Private Sub Quit_Click()
stopped = True
End Sub

Sub Adc10()                        'Get four ten-bit values
Dim i As Integer
Command bits10
For i = 0 To 3
   get_bytes 2                 'next line scales to range
                               '-1 to 1
   Adc(i) = (256! * (bytes_in(1) And 3) + bytes_in(0)) /
   512! - 1
Next
End Sub

Sub Adc8()                         'Get four eight bit
                                   'values
Dim i As Integer
Command bits8
For i = 0 To 3
   get_bytes 1
```

```
   Adc(i) = bytes_in(0) / 128! - 1 'scale
Next
End Sub

Sub get_bytes(n As Integer)   'Read from serial port to
                              'bytes_in()
    buf n                     'wait until n bytes
                              'received
    Serial.InputLen = n
    bytes_in() = Serial.Input
End Sub

Sub buf(i As Integer)         'waits for buffer to hold
                              'i bytes
    Dim j As Integer
    Do
      DoEvents
      j = Serial.InBufferCount
    Loop Until j >= i
End Sub

Sub send(a() As Byte)
    Serial.Output = a()
End Sub
```

The alternative to using a language such as Basic or C for the PIC code is to use assembly language. Mark Phythian has provided a sample of the equivalent code for this example, with just a small portion of the code involved:

```
; define variables
adtable EQU 0x39    ; adc result table 8 bytes
chan   EQU 0x41     ; channel no
val    EQU 0x42     ; word size adc result variable
oldb   EQU 0x44     ; last portb value
char   EQU 0x45     ; single character command
command EQU 0x46    ; upper 4 bits of command
n      EQU 0x47     ; lower 4 bits of command
m      EQU 0x48     ; byte size temporary variable
wrd    EQU 0x49     ; word size temporary variable

; Code executes here at start up
       ORG 0x0000    ;Location to put the code
       BCF PCLATH,3
       BCF PCLATH,4
       GOTO start
```

```
      ORG 0x0004       ;Place interrupt code here at
      address 0004
      MOVWF W_TEMP  ; save registers
      SWAPF STATUS,W
      CLRF STATUS
      MOVWF STATUS_TEMP
      CALL ISR              ; call interrupt service
                              routine
      SWAPF STATUS_TEMP,W
      MOVWF STATUS
      SWAPF W_TEMP,F
      SWAPF W_TEMP,W        ; restore registers
      RETFIE                ;return from interrupt
start:
; setup PIC
      BSF STATUS,RP0        ; select page 1
      MOVLW 0x72
      MOVWF 0x0F        ; set internal RC select to 8MHz
      MOVLW 0xC0
      MOVWF 0x06        ; set PORTB 0-5 pins as outputs,
                        ;6 & 7 inputs

; initialise adc
      MOVLW 0xFF
      MOVWF 0x05        ; set portA as input
      MOVLW 0x0F
      MOVWF 0x1B        ; set PORTA pins 0-3 as analog
                        ;inputs
      MOVLW 0x80
      MOVWF 0x1F        ; set 10 bit A/D result format
                        ;right justify ADRESH/L
      BCF STATUS,RP0        ; select page 0
      MOVLW 0xC1
      MOVWF 0x1F        ; set A/D conversion clock to
                        ;internal source,
                        ; turn on adc
      BSF STATUS,RP0        ; select page 1
      MOVLW 0x7F
      MOVWF 0x01        ; enable PORTB weak pullups for
                        ;inputs 6 & 7

; setup UART
      BSF STATUS,RP0        ; select page 1
      BSF 0x0C,5        ; enable UART RX interrupt
```

```
        MOVLW 0x08
        MOVWF SPBRG          ; set UART BAUD rate 57600
        BSF TRISB,2
        BSF TRISB,5          ; set PORTB bits 2 and 5 as
                             ;outputs for UART
        MOVLW 0x24
        MOVWF TXSTA          ; enable Transmitter
                             ; set High BAUD rate select bit
        BCF STATUS,RP0
        MOVLW 0x90
        MOVWF RCSTA          ; enable Serial port,
                             ; continuous enable receiver
        MOVLW 0xC8
        MOVWF 0x0B       ; enable GIE, PEIE and RBIE for
                         ;UART

; initialise variables
        BCF STATUS,RP0       ; select page 0
        CLRF 0x06        ; clear PORTB
        MOVLW 0xC0
        ANDWF 0x06,W
        MOVWF oldb       ; last value of PORTB bits 6 & 7
                         ;for change of state
; Send "OK"              ; start up ok
        MOVLW "O"
        CALL TXD
        MOVLW "K"
        CALL TXD
```

This does not yet include the receipt and execution of commands. It is clear that the use of the Basic compiler saves a large amount of effort.

12

Machine Vision

The broad subject of machine vision has many levels of complexity. The simplest is the use of a single photosensitive detector to locate the boundary of a brightness change, so that, for example, a factory vehicle carrying parts can follow the edge of a line painted on the floor using "if it's bright, steer left; if it's dark, steer right."

At the other end of the scale is a high-resolution color vision system in which the computer must recognize some object by its shape or texture, even though it might be partially obscured.

Some of the associated mathematical and computational techniques are concerned with improving the "quality" of the appearance an image, while others relate to extraction of data from the image such as the finding of edges and other features.

12.1 VISION SENSORS

In Chapter 2, we met a hierarchy of optical sensors that can be ranked in order of increasing complexity as follows.

12.1.1 Single-Point, Binary

This is "pair" consisting of a single LED and a single phototransistor:

Essentials of Mechatronics, by John Billingsley

- A *reflective opto switch* to detect a dark mark on a light background or vice versa.
- A *slotted opto switch*, where the sensors are mounted to face each other and indicate when there is an obstruction in the slot.

12.1.2 Single-Point, Analog

A single photocell measuring brightness is a popular sensor for a "Micromouse," a robot finding its way through a maze, where brightness can be used as a crude measure of distance from a wall.

A single sensor can be given the attributes of a linescan device by scanning it, such as with the use of a spinning mirror.

An optically based sensor that has had wide adoption as a "quick fix" aid to navigation is the Sick sensor. This uses just such a spinning mirror to scan with a laser beam. The additional factor is that the beam is pulsed. High-frequency circuitry measures the time of flight of the return journey to and from the point of contact. In this way, a map is obtained of the range from the sensor as measured in the scanning plane.

There is much to criticize with this sensor, mainly because of its serial output format. It was originally designed simply as a safety device to ensure that nobody entered the proximity of a dangerous object such as an industrial robot, so the output of image data was intended as a diagnostic tool. The basic scanning rate is 40 scans per second, with maximum resolution representing samples at quarter-degree intervals. Even at 500kHz baud rate, however, the serial output cannot keep up with the highest scan speed at the highest resolution.

12.1.3 Linescan Devices

These are a linear array of sensors, giving data for one line of an image. As in the fax machine, a two-dimensional image is built up by the object moving past the array.

12.1.4 Framescan Devices

A two-dimensional image is captured in one hit. There may be a single frame of data, relating to a rectangular array of pixels, or a stream of frames constituting a "movie."

12.2 ACQUIRING AN IMAGE

For the single-pixel or the linescan sensor, simple bit-level input will be similar whether the system is built around a single-chip microcontroller or a

PC. It is when we wish to acquire a full two-dimensional image that we are faced with a confusion of choices.

12.2.1 DirectX and VFW

For minimum effort, it is easy to purchase a low-cost Webcam and plug it into a USB port on a PC. The driver software that comes with it will enable you to see moving images on the screen, and freeware packages will let you communicate face-to-face with your friends.

There are many cards on the market that can tune a television signal or receive a "composite video" signal (the yellow socket on the VCR). They "stream" the data onto the computer screen, but again we must break into the entertainment-directed technology if we are to make serious use of the signal.

We must answer the problem of putting image data where you can attack it with analysis software.

Close to the hardware level, the "driver" inputs bytes of data and packs them into an array. It then signals software at the next level to indicate that a frame of data is ready, while data bytes continue to be packed into a second frame. An early standard for using such data is called Video for Windows (VFW). An OCX control for Visual Basic can be written to capture data at this level. Details of such an OCX, including the source code, can be found on the Web at www.essmech.com/12/2/1.htm.

As soon as you place this control in your VB form, you can access its properties and methods. One of these is SnaptoArray (I admit that it has more than two syllables!), which will copy the next frame of image data that arrives into an array that you name. What you do with the image is then up to you.

Vision and other media processes are supported in later versions of Windows by a software suite called DirectX. The software developer's kit can be downloaded free from the Microsoft site—although it is several hundred megabytes in size. It includes DirectShow, which deals specifically with video streams.

The package is designed around "drag and drop" concepts, in which "filters" are linked in a "graph." A handy tool for building such graphs comes with the package. It is called Graphedt.exe.

The filters are unlike any of the filters we have met in the control sections. One example of a filter is a videocamera! The filter appears as a rectangle on the chart. In general, it has input and output "pins" that are notional, not physical.

A "video capture filter" such as a Webcam might have two output pins, CAPTURE and PREVIEW. A right click on one of the pins can show its "properties," the format of the data that can be taken from it. A typical value for a Webcam is "Major type: video—subtype RGB24."

A right click inside the rectangle itself will present the choice of FILTER PROPERTIES. In this case, the choice will open a window in which video source and video format can be set or changed.

The other option when the output pin is right-clicked is RENDER PIN. A second box will appear, with label VIDEO RENDERER, with its input pin connected to the output pin of the Webcam rectangle.

In a control bar above are the green triangle and red square for media RUN and STOP. A click on the RUN icon causes a window to appear with the moving Webcam image in it.

Of course, this is just the tip of the iceberg. There are filters for compressing video, for rendering audio, for interleaving video and audio streams in an "AVI Mux," and a file writer to record your video to disk. These are just a few of the hundred or more filters that are likely to lurk on your machine.

So, how is video captured for analysis? The analysis can be performed without capturing it at all. Instead, the analysis software is written as yet another filter that can accept the incoming video stream in real time and pass on the desired conclusions.

The few simple lines of code that are required to process an image, say, to reduce objects to their edges, have to be "topped and tailed" with a mass of "include" references and other housekeeping. However, a colleague, Mark Dunn, has contributed a template and a "wizard" that have been placed on the Website. These will enable you to construct your own image processing filters. You will also find examples that you can modify for your own purposes.

Provided you do not mind depending on one specific commercial operating system, you will find this a satisfying and rapid way to arrive at machine vision solutions. You may instead prefer to take a "bottom-up" approach.

12.2.2 Video Chips

As USB Webcams have tumbled in price, their cousins have invaded mobile telephones. There is a growing market for video subassembly modules for embedding in consumer products, for both low-resolution "fun" applications and high-resolution cameras.

The computers destined to handle these signals are far removed from the PC. They are single-chip microcontrollers, such as the *reduced instruction set computer* (RISC) ARM series. Nevertheless, once the image has been captured into an array, the analysis procedures they apply are almost identical.

12.3 ANALYZING AN IMAGE

Image data bytes flow at an immense rate, even from a low-resolution camera. In RGB24 format, one byte is used for each of the red, green, and blue components of each pixel. An image of 640×480 pixels will require $640 * 480 * 3$ bytes per frame. There will be 30 frames per second (25 in many countries outside the United States), so the data rate is 27,648,000 bytes per second. Even at a resolution of 320×240, the flow is nearly 7 megabytes per second.

It is clear that an essential feature of analysis must be data reduction.

12.3.1 Data Reduction

For many purposes, each pixel can be reduced to a binary decision, light or dark. A vision guidance project studied small green seedlings on an earthy background, and a decision "soil" or "plant" gave all the image data needed. Immediately the data size is reduced by a factor of 24.

Perhaps the largest reduction can be made by looking at just a subset of the image bits. One project concerned the visual counting of macadamia nuts. They were picked up between the blue-colored bristles of a plastic brush roller. The routine needs only to look at every fifth pixel or so to avoid missing a nut. When a nonblue pixel is found, a more intensive search can be made to locate the outline of the nut with some accuracy. Thus the initial scan only looks at 1 pixel in 25 of the image.

The ultimate data reduction in such projects is to the "answer," maybe statements such as "steer left a little" or "there were 2435 nuts."

It is important to discriminate between processing methods that extract "facts" from the image and processing that will simply change the appearance—or processing that will change the appearance as little as possible, for that matter. Image compression such as is used for digital television is a subject in itself.

A black-and-white image is likely to contain lumps of black pixels and lumps of white ones, rather than a random scattering. An early method of data compression was *run-length* encoding, where each scanline is coded in a form that might represent "23 black, 15 white, 75 black . . . ," and so on.

But the clumping will take place in two dimensions, not just along scanlines, so methods such as LZW allow the data to be reduced in size with no loss of actual information.

The compression of color images presents a different problem. This time, image data must inevitably be lost, since only in cartoons will many adjacent pixels be identical in color, but the aim is to keep the "essence" of the appearance of the image.

One compression method is to use a "palette" of 256 colors and approximate each pixel to one of these. The approximation can be brought a little closer by the use of "dither," the alternation of two colors to get that appears to the eye as something in between.

More effective for photographs is the JPEG technique. The picture data defines the parameters of two-dimensional functions, bounded by coarse rectangular tiles of the image. These fit together to give a smooth high-resolution picture, but detail can be lost and flat areas such as sky can carry "tide marks" of color quantization. The degrees of compression and smoothing can be set as a parameter when compressing the image.

Sequences of movie images offer even further possibilities. Many "codecs" (*co*mpression–*dec*ompression filters) save only the differences between frames to the data stream, so that the background does not need to be repeated.

Every few frames, maybe 15 or so, the entire image is saved so that it is necessary to go back only to this "key frame" to reconstruct a particular image, rather than to the beginning of the film.

To a large extent, image compression is irrelevant as far as our purposes are concerned. Allowing for the problems of data size, we wish to work on an image with as much of the original detail left intact as possible.

Now we have captured a frame of image data, either in an array that we can access using subscripts in a high-level language or in a block of memory into which we construct a pointer to find the pixel we seek.

12.3.2 Smoothing a Binary Image

Whether the image is binary or grayscale, we will want to perform some sort of integration or differentiation on it to achieve a filter (in the control sense). The first operation that we will consider is smoothing, to remove spots and ragged edges.

The simplest way to do this is to consider each square block of 9 pixels. Take their average, and give that new value to the center pixel. If the image is binary, "taking the average" means that if 5 or more of the 9 pixels are light, the new pixel is light, otherwise it is dark.

You can see this in action on the Web at www.essmech.com/12/3/2.htm. Examples have been written in JavaScript, which has a syntax closely resembling that of C. More details of the implementation are given in Chapter 13.

The code that performs the smoothing is

```
function smooth() {
 var m, i, j, k, l;
 for(i = 1;i<cols;i++){                //for each point
   for(j = 1;j<rows;j++) {             //except the edges
      m = 0;                           //clear the total
      for (k = i - 1;k<=i + 1;k++) {
         for (l = j - 1;l<= j + 1;l++) {
            m = m + pic1 [k][l]; //add 9 values in 3x3
                                       //block
         }
      }
      if (m > 4) {                     //If majority are white
         pic2 [i][j] = 1;              //make pixel of pic2
                                       //light
      }else{
         pic2 [i][j] = 0;              //otherwise make it
                                       //dark
      };
   }
 }
}
```

In this example, the smoothing is applied several times and the image settles down to a shape without ragged edges.

When used on a grayscale image, this averaging technique has the effect of blurring the image. We will see another method in action later.

A disadvantage is that some special measures would be needed to process the outside boundaries of the image, since they have a row of neighbors missing.

12.3.3 Finding Edges

To find the outline of an object in the image, we must think in terms of differentiating it.

In the world of discrete samples, or pixels, differentiating becomes "differencing," taking the difference between a value and its neighbor. We can easily edit the code of the last example to be

```
function diffx() {
 var i, j, k, l;
 for(i = 1;i<cols;i++){        //for each point
   for(j = 1;j<rows;j++) {    //except the edges
      if ((pic[i+1][j] - pic[i][j]) > 0) {
         //If the pixel to the right is brighter
         pic2 [i][j] = 1;     //make pixel of pic2 light
      }else{
         pic2 [i][j] = 0;     //otherwise make it dark
      };
   }
 }
}
```

So, what does it do? We find a sort of negative shadow, where the lefthand edge is outlined in bright pixels while the rest are dark. This is certainly differentiating the image, but not in a way that is generally useful.

To try it for yourself, open the previous "smoothing" example and copy and paste this new function into the code window below the "smooth" function. Then add

```
diffx();
pic1=pic2;
showpic1();
```

below the rest of the code and it is ready to run.

We could embroider the code to replace the "greater than" sign ">" with a "not equal" sign "<>" to get shadows on both left and right edges, but we would also have to OR a test on the vertical difference if we wish to have an outline all round the object. But there is a more methodical way to do it.

We can think in terms of *convolution*, a process where one array of values is applied to filter another by multiplying and adding corresponding elements, planting a result, and then moving the pointer of the first array to the next pixel.

In this differencing case, the array of the filter is just [–1,1], or perhaps we should write it as [0,–1,1] since it is then clear that the "result" must be written to the central pixel position.

So, if we start with a row of pixels

```
0 0 0 0 1 0 1 1 1 1 1 0 0 0
```

and apply the filter

```
[ 0 -1 1]
```

we will first get the calculation

```
0 0 0 1 -1 1 0 0 0 0 -1 0 0
```

which results in a new string of pixels

```
0 0 0 1 0 1 0 0 0 0 0 0 0
```

when we set anything less than 1 to be a zero pixel.

Each row of pixels will be processed independently to give the full image.

If we want to detect both sides of the object, however, we should instead be looking at the second derivative, or the second difference. Now, convolving the filter with itself, we get

```
[-1 2 -1]
```

meaning "twice this pixel, minus the left and right neighbors."

When we apply it to

```
0 0 0 0 1 0 1 1 1 1 1 0 0 0
```

we get numbers

```
0 0 0 -1 2 -2 1 0 0 0 1 -1 0 0
```

which become pixels

```
0 0 0 0 1 0 1 0 0 0 1 0 0 0
```

We have succeeded in finding the edges, plus the isolated pixel "speckle." The new image is not shifted to the right or left, as it would be if using the previous filter.

So, can this convolution method help us to process the image in two dimensions?

12.3.4 Convolution and Array Filters

Consider the filter

```
 0 -1  0
-1  4 -1
 0 -1  0
```

This is the sum of the previous filter, padded out with a row of zeros top and bottom, added to its vertical counterpart. We can look at it pragmatically to deduce that the new pixel will be light only if the present pixel is light and not surrounded top and bottom, left and right by other bright pixels.

We must define our coefficients to be an array

```
filt=new Array(3);
filt[0]=new Array( 0,-1, 0);
filt[1]=new Array(-1, 4,-1);
filt[2]=new Array( 0,-1,-0);
```

and these coefficients can then be used in the routine

```
function filter() {
var m, i, j, k, l
  for(i = 1;i<cols;i++){          //for each point
    for(j = 1;j<rows;j++) {  //except edge
       m = 0;
       for (k = 0;k<=2;k++) {
          for (l = 0;l<= 2;l++) {
             m = m + pic1 [i+k-1][j+l-1]*filt[k][l];
          }
       }
       if (m >=1) {               //If total is positive
          pic2 [i][j] = 1;        //set pixel of pic2 to red
       }else{
          pic2 [i][j] = 0;        //set pixel of pic2 to black
       };
    Label(m,i,j);
    }
  }
}
```

See it in action at www.essmech.com/12/3/4.htm.

A host of filters are based on the use of such a 3 × 3 array of coefficients. We could have used this technique for the first smoothing example by defining the array to be

```
0.2 0.2 0.2
0.2 0.2 0.2
0.2 0.2 0.2
```

Only if 5 pixels in the array of 9 are bright will the total reach the value that we have set for the threshold. We could indeed try other values than 0.2. The value 0.25 would let us set the criterion at 4 bright pixels.

But returning to edge finders, our criterion could be that a bright pixel would survive as an edge unless surrounded completely by other bright pixels. The filter would be

```
-1  -1  -1
-1   8  -1
-1  -1  -1
```

Perhaps we want the edge to be marked in the "sea" that surrounds the "island":

```
1   1  1
1  -8  1
1   1  1
```

The choices are endless. You can try out any you think of on the Web.

I must repeat that these operations will change the appearance of an image, reducing it to spots with the appearance of lines at the edges of any "blobs," but a lot more has to be done before a line can considered as a "path" around the object.

12.3.5 Smoothing Grayscale Images

To see an array filter in action on a grayscale image, see the first Web example at www.essmech.com/12/3/5.htm.

The filter in this case is

```
 1,  -2,   1
-2,   4,  -2
 1,  -2,   1
```

giving the result that would be obtained if first the horizontal "second difference" operator [–1, 2, –1] were applied, followed by its vertical counterpart.

It is clear that by taking a second difference, it has eliminated the "graded" background that would give problems if a single threshold had to be set.

In control terms, the convolution that we are performing is termed a *finite impulse response* filter. The distance of the influence of any sample is limited, in this case to its neighbor. Broader filters could be set up, 5 × 5 or maybe 7 × 7, but the computational effort increases with the square of their size. Alternatively smaller filters could be used repetitively, so that the influence will "spread out" one pixel at a time.

There is another approach. In control theory, we saw that a lowpass filter (see, e.g., Fig. 12.1) would smooth a time series. Such a filter might take the form

```
for i = 0 to n
  xslow = xslow + (x(i) -xslow) / k
  x(i) = xslow
next i
```

where k determines the time constant.

But this "smears" the waveform to the right and would similarly smear an image. In real time, we can process a time series in only one direction, but here we have a captive image. We can follow up the left-to-right smoothing with another smoothing right-to-left that will exactly cancel out the smearing, while leaving the blurring in place. This approach, using a two-way filter, is shown in Figure 12.2.

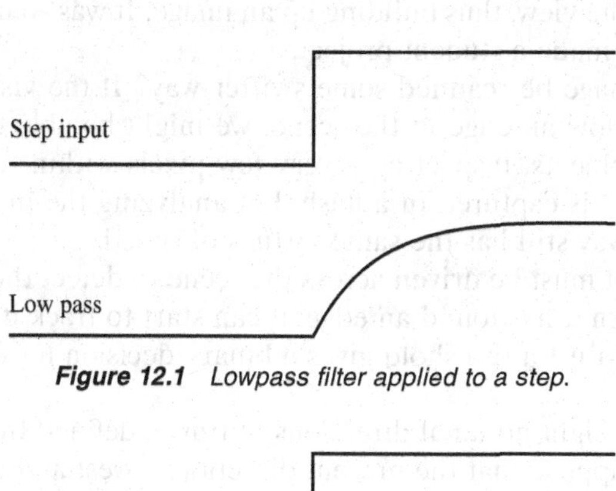

Figure 12.1 *Lowpass filter applied to a step.*

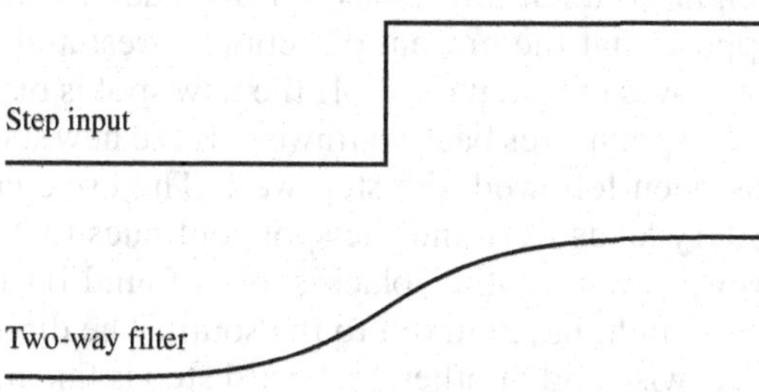

Figure 12.2 *Two-way filter applied to a step.*

Having blurred the image horizontally, we get the final effect by applying a similar vertical filter. See the result in the second Web example at www.essmech.com/12/3/5.htm.

12.3.6 Sharpening a Grayscale Image

In control theory, we used a *highpass* filter to approximate to differentiating a signal. We saw that we could construct such a filter by first making a lowpass filter, then subtracting the smoothed version from the original signal. The same principle applies to sharpening an image.

We have just seen how to use an *infinite impulse response* filter, our simple lowpass filter, for smoothing an image. We apply it left-to-right, right-to-left, top-to-bottom, and finally bottom-to-top.

When we subtract the smoothed version from the original image, we have a sharpened image in which the edges are enhanced. If this difference image requires more contrast, we can multiply the values to enhance it.

See the images shown in Figure 12.3 in action on the Web page at www.essmech.com/12/3/6.htm.

12.3.7 Edge Tracing

In the mid-1980s, we had attached a primitive stepper motor robot to a computer. A lens and a simple photocell were then added to the gripper of the robot to give a single point of vision. The robot could be moved to scan the photocell over the view, thus building up an image. It was somewhat slow and unwieldy, but it made a student project.

Could the image be scanned some swifter way? If the vision point could be driven to follow an edge in the scene, we might be able to trace out the boundaries of objects, inspecting a very few pixels within the whole scene. Today the image is captured in a flash, but analyzing the image in a logical and economic way still has the same virtues of speed.

First, the spot must be driven across the scene to detect the first change in brightness. When it has found an edge, it can start to track it. A comparison of brightness against a threshold gives a binary decision for each spot, black or white.

The spot has eight notional directions of travel, defined by eight points of the compass. Suppose that the present direction is west and the present spot is white. The spot moves one step north. If the new spot is black, the edge has been crossed, so the spot moves back southwest. If the new spot is white again, the boundary has been followed one step west. The cycle can repeat for as long as the boundary leads west and the spot continues to "stitch" along it.

Suppose, however, then another black spot is found on the "back step". Then the boundary might have curved to the south. The direction of travel is turned 45° anticlockwise and another backward step is taken, now due south.

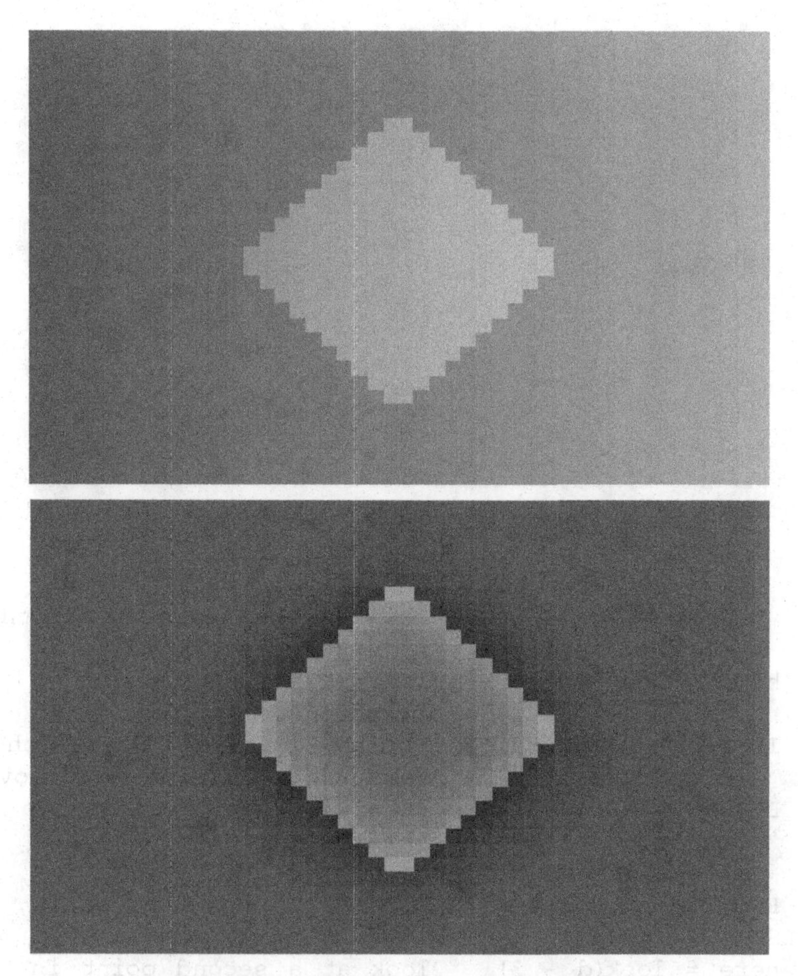

Figure 12.3 Screen grab of edge enhancement.

If there is a change to white, the stitching can continue, now in the new direction; otherwise yet another anticlockwise turn and backward step are taken. Eventually the movement must find an edge again, even if it has to complete the semicircle back to a previous point.

Similarly, if a forward step fails to find a change, the turn is clockwise and another step is taken.

Each time a change is detected, the coordinates are noted, giving a sequence of points that track in order around the boundary. In this case, the path will track anticlockwise around a white object, or clockwise around a black one.

An image created by edge stitching is shown in Figure 12.4.

The outline of the algorithm in QBasic is as follows:

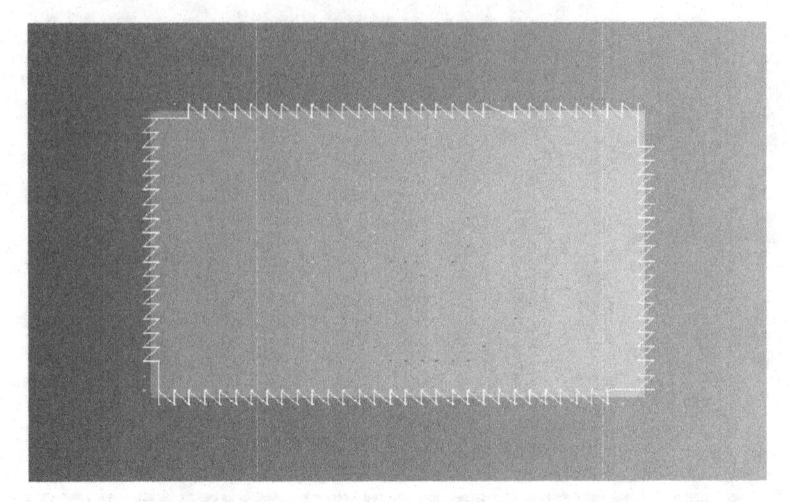

Figure 12.4 *Illustration of edge stitching.*

```
DO                            'This is the search algorithm
 IF here = white THEN         'if first point is white
   here = look(d)             'look at a second point in
                              direction d
   IF here = white THEN       'if it's the same, white, then
     turn 1                   'turn clockwise for next move
   ELSE                       'otherwise you've crossed a
                              'threshold
     notepoint                'so mark it
   END IF
 ELSE                         'if first point was black,
   here = look(d - 3)         'look at a second point in
                              'direction d-3
   IF here = black THEN       'if it's the same, black, then
     turn -1                  'turn anticlockwise
   ELSE                       'otherwise you've crossed a
                              'threshold
     notepoint                'so mark it
   END IF
 END IF
LOOP UNTIL beenthere > 0      'keep going until you hit an
                              'old marked point
```

From this fundamental principle, a number of additions are needed to make the routine work.

If the starting point is not near an edge, the routine will just go round in small circles forever. The first modification is to count the number of steps since the last boundary crossing. If this exceeds four, the direction is not allowed to change until after two steps, then three, and so on. The length of a "straight" increases by one every eight steps. Now, after the first semicircle, the search expands in a spiral.

The second modification enables the program to adapt the threshold to find subtle shades. Two variables hold the lightest and the darkest values found so far. The threshold level is set midway between these levels.

To adapt to local changes, the "lightest" value is reduced by a small amount at each step while the "darkest" is increased. They will ramp until they hit the values being found locally, while variations in the level will keep them apart. If the "gap" becomes too small, boundary crossings are ignored, so that the search spirals out to find a more prominent edge. If the search arrives at a point already tagged, then it is ended.

That describes the details of the technique, but what does it achieve?

The boundary is revealed as an ordered sequence of points, forming a *Freeman chain*.

12.3.8 Analyzing Boundaries

The sequence of points found by the "stitching" method can equally be regarded as a chain of vectors joining one point to the next. Each vector has a length and a direction. If the vectors are added in pairs or more, the "ragged" nature of an oblique edge will be smoothed.

We can take the sum of the lengths of the vectors from the starting point to obtain the distance s moved around the perimeter, and against this we can plot the angle ψ of the vector, the direction of the perimeter at that point, to get an $s-\psi$ (s-psi) curve.

With this curve the shape data can be reduced to a few hundred bytes of data, say, 256 or 512, representing the tangent directions at equal intervals around the perimeter. By comparing this shape data against templates of the same length, we can recognize the shape.

Object recognition might at first seem a daunting task. Even if the size of the object is known, it can be rotated to any angle and be located anywhere in the picture. If you are searching in 0.5 mbyte of data by a correlation method, the number of computing operations is huge. Given an $s-\psi$ plot, most of the task is already done.

From start to finish the angle will change by 2π. This plot will be the same, wherever the object is in the picture. However, the object could be lying at a different angle. In this case the plot will still be the same, if regarded as a cyclic function, but will have a constant added to the angle value. If the object is "flipped," the function will be reversed.

In each case, the task of matching the unknown object against a template is a simple case of examining a few hundred data points. We still have to consider the match as a correlation, shifting the starting point around the template, unless we can find a strategy for determining a starting point. Even so, the computing load is relatively modest.

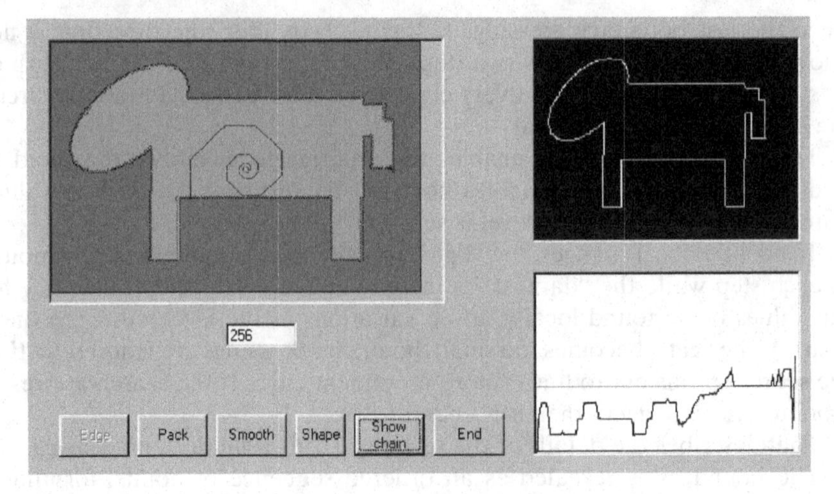

Figure 12.5 *Boundary tracing and s–ψ curve.*

Since ψ is plotted against the proportion of the distance traveled around the perimeter, size does not matter. Objects of the same shape will have the same data, however big they are.

An example on the Website shows a shape being traced. The $s-\psi$ curve is then generated, smoothed, and reduced to a fixed length. At each stage, the shape is reconstructed so that it can be seen how far the smoothing might distort the shape.

See http://www.EssMech.com/12/3/8.htm and Figure 12.5.

Of course, clues other than shape can be used for a comparison, once the binary decision has been made to discriminate between the pixels of the object and the background:

- The area of the object can be found by counting pixels.
- By searching for boundaries within the object, any "holes" can be counted.
- By testing the "width" of the object as it is rotated, the ratio between maximum and minimum can be found.

This is not the only format for shape data. By taking moments, the center of gravity can be found. Now, by tracing out along radii from the center of gravity, radius length can be found as a function of radius angle. However, a curve may have a reentrant "hook" so that a radius can cut it in more than one place. The plot of radius versus angle is then no longer single-valued and is therefore difficult to list as a computer function.

This is just a glimpse of the vast range of possibilities that are opening up in vision sensing. Any more would go beyond the essentials.

13

Case Studies

13.1 ROBOCOW—A MOBILE ROBOT FOR TRAINING HORSES

The National Centre for Engineering in Agriculture received a startling project proposal concerning a "robot cow" that could fool a horse. As discussions progressed, it rapidly became clear that the design requirements would be extremely hard to meet. The purpose was the training of horses for cutting contests, where horse and rider must control the movement of a young cow. The business proposal was made by an acknowledged "cutting champion."

In heading off a young cow that is trying to rejoin the herd, the partnership of horse and rider depends entirely on the ability of the horse to recognize and anticipate the intentions of the cow. In training the horse, it has been usual to use a "borrowed" calf—or two, or three or more.

As fast as the horse learns, so the cow also learns and very soon refuses to cooperate. So, to train one horse takes the use of many cows and considerable expense. A robot cow, on the other hand, would be consistent and predictable by the rider if not by the horse.

Robocow, as it was quickly named, must perform a memorized sequence of actions, so that with no more than a two-button controller the horse rider can select, start, pause, or resume the cow's performance. In addition, it is important that when completing a sequence designed to bring it to the starting

Essentials of Mechatronics, by John Billingsley

point, the cow can repeat the sequence several times before large positional errors are built up.

Some of the sequences can be preprogrammed during manufacture, but there is also the need to provide the ability to memorize special individualized routines on the farm. For this, a standard radio-control joystick system is used.

At the outset, the mechanical performance requirements were challenging, to say the least. The cow must reach a speed of 20 km/h with an acceleration of several meters per second per second. The terrain was specified as "beaten earth."

13.1.1 Overview

The selected geometry was a steered tricycle with driven front wheel—the same system as the fairground "dodgem." The steering can turn through half a circle, so that the cow can spin about the center of its rear axle. It can actually accelerate faster in reverse, when the weight is thrown onto the driven wheel.

Navigation of Robocow (Fig. 13.1) depends on odometry. The undriven rear wheels are equipped with Hall effect sensors that enable their angles to be monitored at all times. Heading is deduced from the difference between the wheel rotations, and the coordinates are estimated by integrating forward motion multiplied respectively by the sine and cosine of the heading.

Steering control uses a highly nonlinear algorithm that drives the system to a new setting in a fraction of a second, very similar to the position control discussed in Chapter 3. The overall result is a lively beast where performance is limited more by skidding on the dirt surface than by any limitations in the drives or controls.

13.1.2 Mechanical Design Considerations

Since the drive is applied through the front wheel, the two rear wheels are undriven and have no reason to slip in the direction of progress. They can therefore be used for reliable odometry, provided they do not leave the ground.

A single 120 W motor did not give the lively action the clients were seeking. With two such motors mounted on the front-wheel assembly, symmetrically placed fore and aft of the wheel, there was more than enough drive to skid the front wheel.

The steering uses another very substantial motor of 60 W rating, so that it can be driven from one extreme to settle at the other in well under one second.

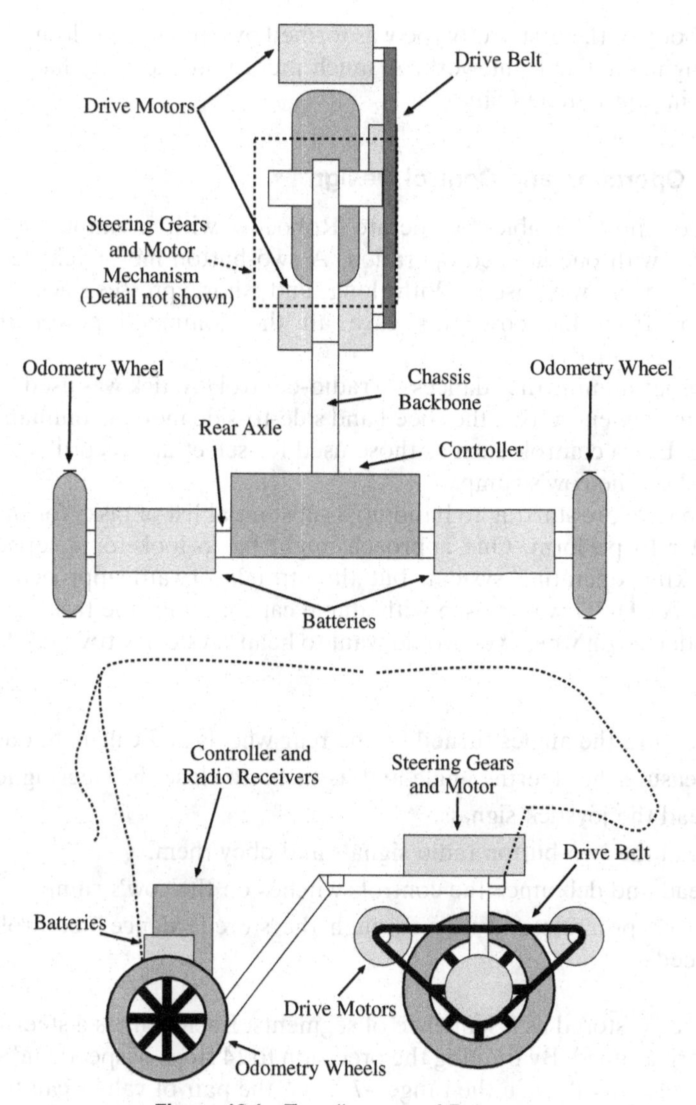

Figure 13.1 *Two diagrams of Robocow.*

Two 12 V lead acid batteries are mounted symmetrically near the rear wheels. There is a tradeoff between wishing to keep the center of gravity over the driven wheel and the need to keep it aft to lessen the risk of rolling.

The diameter of the front wheel is 310 mm and of the rear wheels is 300 mm. The wheelbase is 585 mm fore and aft, while the rear wheels are 760 mm apart.

The body of the first prototype was formed by stretching a cloth "cow suit" over a light tubular framework. A much more realistic body has now been molded in polystyrene foam.

13.1.3 Operation and Control Design

The rider must be able to operate Robocow with a simple pushbutton controller with one-handed operation. A two-button motor-vehicle remote-locking device was used. With long and short presses, acknowledged by beeps from the cow, this gave all the command power that was needed.

For programming the "dances," a radio-control joystick was used, with fore and aft movement setting the speed and side-to-side motion commanding the steering. Extra controls such as those used for selecting "record" mode were mounted on the cow's rump.

By now we are starting to build up a substantial list of tasks for the microcontroller to perform. One approach might be to look for a sophisticated multitasking operating system, but the straightforward approach is much simpler. An HC11 was chosen with ample capability for the task itself, but it is not a device on which you would want to heap "system software." The tasks are

- Measure the angles turned by the rear wheels and calculate odometry.
- Measure the steering angle and its tacho to close the steering loop.
- Read the joystick signals.
- Read the two-button radio signals and obey them.
- Read and debounce the control switches on the cow's rump.
- Check progress and step through the stored "dance," controlling the speed.

The dance is stored as a sequence of segments. Each defines a steering angle and a target speed. By limiting the precision to 14 steps of speed and steering, represented by values in the range –7 to +7, the pair of values can be held in a single byte. For each segment there is also stored a termination condition in a second byte.

If the segment is a turning one, the termination condition defines the heading angle at which the segment ends. If the steering is required to be straight in the segment, its termination condition defines the aggregate distance to be covered until the next segment starts. Segments are short enough that using a single byte will identify the least-significant byte of the distance without ambiguity. If the speed command is zero, the condition determines the time that must elapse before continuing.

The termination conditions are absolute to avoid accumulating errors; that is, the distance termination is not simply the length of the segment but the

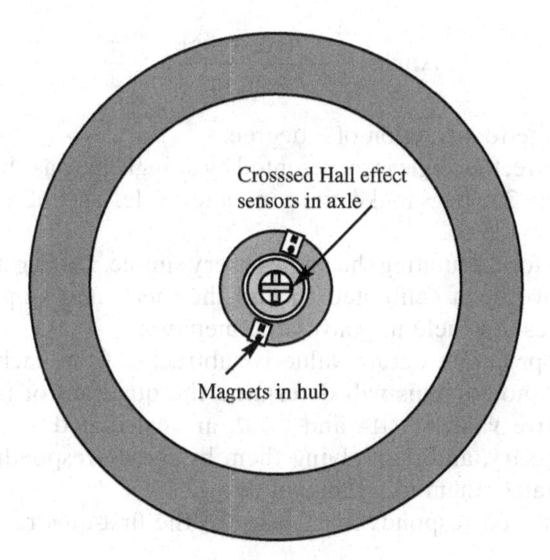

Figure 13.2 *Hall effect sensor, shown in a wheel.*

total distance covered since starting. Similarly, the heading is absolute, measured from the start condition in terms of the difference between left and right wheel rotations.

When required to run straight, any error in steering calibration could cause the path to be a large circle. This is avoided by applying heading error feedback to the steering, nudging it by an angle limited to a few degrees. It keeps the machine straight without being perceptible in its behavior.

13.1.4 Sensors and Control Loops

At the heart of both odometry and steering sensors is the simple Hall effect angle-sensing device mentioned in Chapter 2. Two UGN3504 analog magnetic sensors are mounted with their sense axes perpendicular to the axis of rotation and perpendicular to each other. In the case of the wheels, they are mounted within the rod that forms the wheel bearing. A magnet is mounted on the wheel with its axis radial, normal to the rotation axis (see Fig. 13.2).

The sensors therefore give signals representing the sine and cosine of the wheel rotation angle. In fact, two such magnets are used, mounted on either side of the axis, so that second-harmonic distortion is minimized if the sensors are not exactly axial.

The sensor signals are encoded directly by the 8-bit ADC channels of the 68HC11 microcomputer that controls the cow. A simple and novel routine extracts the angle from these two signals.

The routine is based on the approximation

$$\text{Angle} = \left(\frac{3.7(\sin - \cos)}{2.7 + \sin + \cos} + 1 \right) \frac{\pi}{4}$$

which is accurate to a fraction of a degree.

In the software, the angle is represented by a single byte as "binary degrees" or "begs," where 256 begs make one complete circle. The increments are thus slightly less than 1.5°.

The routine for calculating the angle is very simple. During an initial setup, the sensors have been calibrated to find the mean and amplitude of their variation, values now held in nonvolatile memory.

First the appropriate datum value is subtracted from each sensor signal and the sign is noted. This will determine the quadrant of the final result. Now the positive values SINA and COSA are calculated by negating these values, if necessary, and multiplying them by the corresponding calibration factor to normalize them to a range of 0 to 127.

The value $\pi/4$ corresponds to 32 begs, so the first-quadrant angle is now given by

```
237*(SINA-COSA)/(346+SINA+COSA)/2  +  32
```

since $2.7 * 128 = 346$ and $64 * 3.7 = 237$

The wheel angle is "extended" into a multibyte value that is long enough to hold the total number of wheel revolutions for a performance. If the new "local" value is within a count of 64 of the previous value, it is clear whether a carry or a borrow should be propagated into the higher bytes. If the change is greater than a quarter of a revolution, an error is indicated.

The alternative to this analog technique would be to use a bidirectional counter to count pulses from an encoder disk. In that case, unless a hardware counter is used, the service routine would have to interrogate the transducer more than 256 times per revolution of the wheel. With the analog sensor, four or more interrogations per revolution will be sufficient.

In the wheel angle routine, a highpass routine with which you will now be familiar gives an estimate of the speed of each wheel.

The steering sensor has an identical pair of "crossed Hall effect" sensors and uses the same subroutine to calculate the angle. The crispness of the steering is made possible by the addition of an analog tacho. This signal is encoded by an additional ADC channel, bringing the total of channels to encode to seven.

13.1.5 Software Structure and Timing

The software framework was designed by Jason Stone, of the NCEA.

After initialization, the software enters an "idle loop" waiting for commands from the two-button switch or the rump switches. Whether recording or playing a dance, the routine also follows a simple loop. All the control is applied within interrupt routines.

A timer interrupt initiates a control cycle every 4 ms, which starts by reading the sensors and updating the odometry. The steering is serviced and a counter selects every tenth interrupt, so that speed control and the keypad routines are serviced every 40 ms.

Every 4 ms the steering angle is measured and a required steering velocity is computed. This velocity demand is limited in magnitude. The measured steering velocity is added and the result controls the bang-bang (with dead-zone) drive to the motor. The deadband is only one "beg" when the cow is moving or when the demanded angle is nonzero, but is increased when at rest to conserve battery life.

The other two important sensor channels are the joystick controls. A conventional model aircraft radio system is used, in which the commanded values are represented by a variable pulsewidth. These pulses are received at times that the software cannot "expect." Their widths must therefore be measured in another interrupt routine, where they are "parked" for processing every 40 ms.

Speed control is applied only every 40 ms, unless there has been a change in command. The velocity is, however, computed every 4 ms.

The lower 2 bytes of the multiturn wheel angle are used as an input to a numerical lowpass filter, with time constant 64 ms, and the difference between actual and filtered values will represent an actual velocity. The quantization level of this velocity is one sixteenth m/s. It is divided by 8 to give a quantized speed zone in the range ±7, where each unit is 0.5 m/s and the top controlled speed is 3 m/s. If the demanded speed is +7 or –7, continuous full power is applied to the motors.

If the measured speed is less than the demanded speed, 12 V of acceleration is applied to the motors. If the speed codes are equal, or if the measured speed exceeds demand by one unit, the motors freewheel. If the speed exceeds demand by two or more counts, braking drive is applied. Speed zero is deemed to belong to the reverse direction set, so that when moving forward and commanded to stop, braking will continue until the zero-speed zone is reached.

13.1.6 In Conclusion

There is some video of Robocow in action on the Web at www.essmech.com/13/1/6. An early prototype was seen on UK television in a *Tomorrow's World* program, while another prototype was placed on display in the Powerhouse Museum in Sydney.

13.2 VISION GUIDANCE FOR TRACTORS

Some years ago, a vision guidance system was developed at the University of Southern Queensland to the stage where commercial exploitation was attempted. Six prototypes were tested by farmers in Australia, and two more were put on trial in the United States. Over the years of the project, there were several changes of imaging technology but the fundamental principles remained consistent. New funding has seen a rebirth of the project, now to be integrated with GPS guidance.

13.2.1 Introduction

The system derives its guidance signal from a videocamera image of the rows of a crop ahead of it, such as cotton. The patented strategy makes it relatively insensitive to additional visual "noise" from weeds, while by tracking several rows at a time it can tolerate the fading out of one or more rows in a barren patch of the field. The image of each row is tested for "quality."

Experimental results showed that the system was capable of maintaining an accuracy of 2 cm. Farmer responses from the extensive field trials were full of enthusiasm—but they still did not purchase the system in sufficient quantities to keep it alive.

The need for automatic guidance of farm vehicles had been recognized for a considerable time. Many guidance methods had been considered, ranging from buried leader cables to beacons, surveying instruments, or satellite navigation. GPS was in its infancy at the time of the original project. All had their drawbacks. The most appealing method was to follow human practice and take guidance from the crop itself, steering the vehicle by means of the view of the rows ahead.

Consistent accuracy of row following allows cultivator blades to be set much closer to the plants, greatly increasing the efficiency of weed control and circumventing the need for additional spraying. Meanwhile, the driver can give greater attention to the cultivation operation and the state of the crop.

But enough of the sales talk. How does it work?

13.2.2 Design Tasks

The design presented a succession of problems:

1. Acquiring an image
2. Determining what pixels represented "plant" and which were soil
3. Separating each row from the others
4. Fitting a line to the center of each row
5. Analyzing slopes and intersections to find a vanishing point
6. Deducing turning of the tractor by using movement of the vanishing point

7. Deducing position error of the tractor from the changes in slopes of the fitted lines
8. Constructing a steering signal

And that is just the vision part of the problem. Then we had the tasks of

9. Adding hydraulic valves to actuate the steering
10. Measuring the steering angle for feedback
11. Closing the steering loop
12. Designing the overall feedback loop that applies the vision-based demand

This last stage is far from trivial. For safety, the steering loop had a slew rate limiter. This was in the form of a simple oil-flow restrictor. In the event of a malfunction the tractor steering could not suddenly slew and cause the tractor to roll over. But the introduction of this nonlinearity brings some severe control problems.

A simulation at www.essmech.com/13/2/2.htm shows that while a small disturbance might be corrected quickly and easily, a larger disturbance can send the same system into oscillation. As we will see later, the control must be designed with nonlinearity in mind.

13.2.3 Image Acquisition

The early work was based on vision systems with very limited capabilities. Far from hampering the project, these limitations almost certainly contributed to its success. It is my opinion that other researchers were led astray by an excess of data and that problems were tackled that did not really relate directly to the fundamental task of steering.

The first of our experiments used a binary "frame grabber" that yielded a black-and-white image—no gray levels—with a resolution of 768 horizontal points by 96 rows vertically.

The image transfer was performed by direct memory access (DMA) to be captured in an array in main memory. Here the software was able to access it for processing. At some cost in overall speed, part of the image was intermittently copied directly to the display memory so that it could be seen on the computer screen and the effectiveness of the algorithm could be assessed by eye. Only a decade before the time of writing computers were much slower.

A later version used a camera interface developed for the consumer market, the "Video Blaster"—marketed in numerous revisions. A full-color image was captured in the onboard memory, and this image could be merged "live" in a window forming part of the display.

The system did have attendant disadvantages, of course. The image memory was mapped at a high address in extended memory, usually selected to be at 15 Mbytes. (That was high in those days!) Occupying 0.75 Mbyte of addressing space, modest computer speeds meant that care had to be taken to select only a small proportion of the data.

With the availability of color, better discrimination was achieved. A field with a newly shooting crop may be littered with light-colored detritus that makes it difficult to discern the crop rows if brightness alone is used. The use of a green filter over the lens provides no improvement. With color, however, it was possible to use the chrominance signal rather than luminance to acquire an image based on the "greenness" of each point.

Today we have a stream of Webcam data with 3 bytes for each pixel, representing red, green, and blue. "Green minus red" is one combination that will give a signal that depends on color, rather than brightness.

Commonality between the evolving hardware versions has been achieved by the use of a function, `picbit(x,y)`, which presents the image in a standard form to the analyzer irrespective of the system from which it is acquired.

13.2.4 Separating Plant from Soil

The level (whether brightness or resolved color component) of the image is now held as a two-dimensional array of values. The first task is to discriminate between the crop and the background field, something made harder by clouds that can change the light levels from moment to moment.

Other researchers had devoted a nine-page paper to this discrimination problem. They argued that the pixel values could be clumped into two separate peaks, corresponding to plant and soil. With the "leafiness" of plants and the lumpiness of soil, this did not seem to have always been the case.

We found a much simpler approach to be successful. From the state of the crop we know roughly the proportion of the ground that is covered. As we pick pixels to analyze, we keep count of the numbers that are reported respectively as plant and as soil. If their ratio is higher than the expected groundcover ratio, we increase the threshold by one step; if it is less, we decrease the threshold. It is as simple as that.

Within a few frames the row images are seen to "fatten up" to match our density expectations. If they do not match the view from the cab window, a tap on a button will change the density parameter until they do. The simplicity of this level adjustment strategy is a heritage of the original binary frame grabber, which made a more complicated strategy unreasonable.

13.2.5 Separating the Row Images

We used a simple technique that depends somewhat on a circular argument. If we know where the rows are, we can define "keyholes" in the image so that the pixels of any keyhole will contain only the image of one row, plus some soil either side of it.

Now our task is reduced to one of finding how the keyhole should be moved to the center of the row within it. Since this involves only calculation, rather than steering movement of the tractor, the correction can be applied to the very next frame to track the rows as they appear to move about.

That still leaves the problem of finding the rows in the first place. But when the tractor is driven straight, we know when to find them. When the quality of fit is insufficient, the windows drift back to the central "straight-ahead" position. Only when it has a good lock on the rows can the system signal that controls automatic steering be engaged.

13.2.6 Fitting Lines to the Rows

The condition of the crop changes through the growing cycle. Initially the plants appear as rows of small dots among other scattered random dots that are weeds. Later they fuse to form a clear solid line. Before long, however, the lines have thickened and threaten to block the laneways. Great tolerance in the vision algorithm is thus required to fulfill all the seasonal requirements.

Figures 13.3a–13.3f are slides from an early presentation, showing how it was done.

The lines tilt either side of vertical, so it is logical to use the form

$$x = ay + b$$

to describe them. The horizontal distance of a point from the line is

$$x - ay - b$$

To fit these to the rows, once again simplicity is the order of the day. It would be a mistake to attempt to analyze the shape of the row boundaries, especially in the early stages of growth. Instead, the "plant" pixels can be treated as points on a graph, through which a straight line is to be fitted.

The regression method is used to fit the best straight line to a set of points. The regression line minimizes a quadratic cost function, the sum of the weights of the points times the squares of their distances from the line. This cost function can be thought of as similar to the moment of inertia of the data points, represented as masses corresponding to pixel values, when spun about the best-fit line.

In our case, we are interested in the horizontal distance from the line, rather than the perpendicular distance, so that the cost function becomes

$$C = \sum_{x,y \in \text{keyhole}} m(x, y)(x - ay - b)^2$$

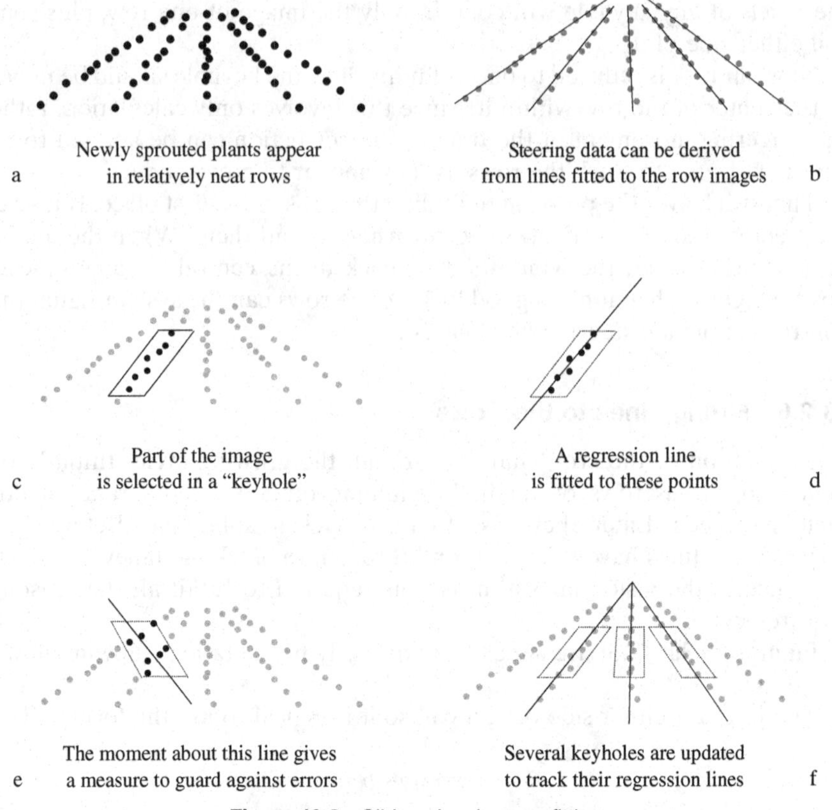

a
Newly sprouted plants appear
in relatively neat rows
Steering data can be derived
from lines fitted to the row images
b
c
Part of the image
is selected in a "keyhole"
A regression line
is fitted to these points
d
e
The moment about this line gives
a measure to guard against errors
Several keyholes are updated
to track their regression lines
f

Figure 13.3 *Slides showing row fitting.*

(This is actually a double summation, since we must sum over both x and y.) Now we want to find the values of a and b that will minimize C. At this combination of values, the partial derivatives of C with respect to a and b will be zero:

$$\frac{\partial C}{\partial a} = 0, \quad \frac{\partial C}{\partial b} = 0$$

When we differentiate, we are still left with the summation giving two simultaneous equations in a and b, involving coefficients that are the following sums:

$$\sum m(x, y), \quad \sum xm(x, y), \quad \sum ym(x, y), \quad \sum xym(xy) \quad \text{and} \quad \sum y^2 m(xy)$$

which in our code we will call

```
m, mx, my, mxy, myy
```

Instead of a and b, we give the results the more descriptive names `fitx` and `fitslope`, so that the solution is calculated by

```
fitx = (mx * myy - mxy * my) / (m * myy - my * my)
```

and

```
fitslope = (m * mxy - mx * my) / (m * myy - my * my)
```

The results are delivered to the steering process in the form of the lateral movement of the vanishing point and the slope of the rows in the picture. From these we can calculate the lateral displacement at any distance in the rows ahead.

If we also calculate the value `mxx`, we can find the actual minimum value of C. If the fit is good, the result should be small. If the crop is scattered or confused with weeds, however, the moment of inertia will be larger. As a test, the minimized value of C is divided into `m` times the moment that we would get if every point were plant, to give a measure of `quality`. The steering information is acted on only if `quality` is sufficiently high.

Often a row may fade out halfway down the field. For this reason, the computation is performed not just for one row but for two or for three. (If all rows are found to be unacceptable, 3 times in succession, then an alarm sounds and control reverts to manual.) Finally, the mean value of all the samples in the keyhole is used to adjust the brightness or greenness threshold for the next frame.

13.2.7 Applying Steering

The main steering tasks were outlined above in items 9–12 in the list in Section 13.2.2. First we must provide a way of converting the electrical signal into the mechanical steering action. One approach would be to turn the steering wheel by means of a motor, but instead we decided to exploit the hydraulic steering of the Case Magnum tractor that had been lent to us.

Two solenoids that drove a relatively simple valve gave us an action in which the steering could be controlled to slew to the left or right. Constrictors, disks with small holes in them, were added to the valve to limit the rate of slew. If anything went wrong with the electronics or electrics, it was essential that the action not be too violent. The human driver must be able to counteract any such steering action.

Now we needed a measurement of the steering angle, to use as feedback. For this, the Hall effect sensor was ideal; it is mentioned in Section 2.2.1 and described in more detail in Section 3.5.2. We can now calculate the angle

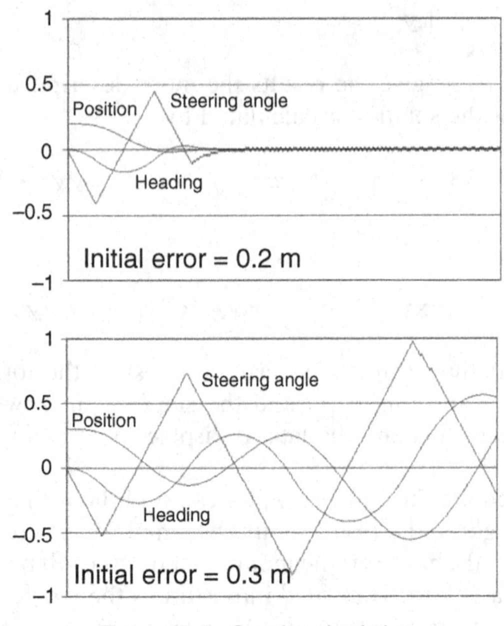

Figure 13.4 *Steering simulation.*

error and simply set one or other solenoid according to the sign of the error, leaving a small deadband in between. The rate limit, although essential for safety, does have a control drawback.

The error signal taken from the vision system is the apparent lateral shift of the rows partway up the picture. Since it is measured ahead of the vehicle, this will have a value that is a sum of the vehicle displacement and a term proportional to heading relative to the row.

In the strategy of a simple linear controller, the demanded steering angle would be made to be proportional to this error. When the rate of change of the steering angle is limited, however, an abrupt onset of limit cycle instability can occur if the initial error exceeds a relatively modest value. This is portrayed in the simulation at http://www.essmech.com/13/2/7.htm and shown in Figure 13.4. The result is to be noted more for its qualitative effect than as an exact prediction of the error magnitude at which instability will break out. Here it shows a nearly ideal response from an error of 0.2 m, while an error of 0.3 m results in disaster.

Of course, the magnitude of the disturbance at which instability occurs can be increased by reducing the steering gain or by choosing a point further ahead from which to take steering data. In either case, however, performance is lost and the response time for recovery from a disturbance is increased.

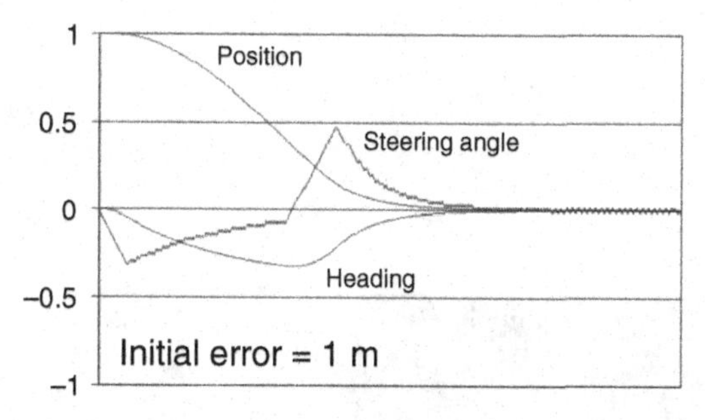

Figure 13.5 *Steering with limited heading demand.*

By rearranging the algorithm in accordance with the topological, nested-loops approach outlined in Section 10.1, we can calculate a succession of demands to which we can apply limits. The simulation mentioned above has been arranged in this form, but the limits have been set too loosely to have any effect. We see a system that is linear apart from the steering rate limitation. You should experiment with the simulation to try various values of limits on steering and heading angles.

Imposing a limit on the heading angle, the amount by which the vanishing point is seen to move, has a dramatic effect as shown in Figure 13.5. Here the limit is set at 50 pixels and an error of one meter is seen to settle with no problem.

Figure 13.5 shows the advantage of the row-fitting approach, identifying the vanishing point movement, over the simpler strategy of inspecting a single horizontal line of the image to measure a displacement.

13.3 A SHAPE RECOGNITION EXAMPLE

As part of a doctorate project, Mark Dunn is working on the discrimination of animal species. Many different animals will approach a watering point in the Australian outback, including the sheep and cattle for which it is intended, kangaroos and other native animals, and also feral species that are regarded as pests.

Feral pigs do untold damage, but feral pork is a commodity that has commercial value. As the animals move past a recognition system, a gate moves to one of two positions, giving access to one of two enclosures. In one of them, animals can reach water and after drinking can exit and go on their way. In the other, feral pigs will gather, drink, and be held for the later attention of the farmer.

The image analysis can be made much easier if the approach has a blue background, such as a tarpaulin, but this might deter the animals from

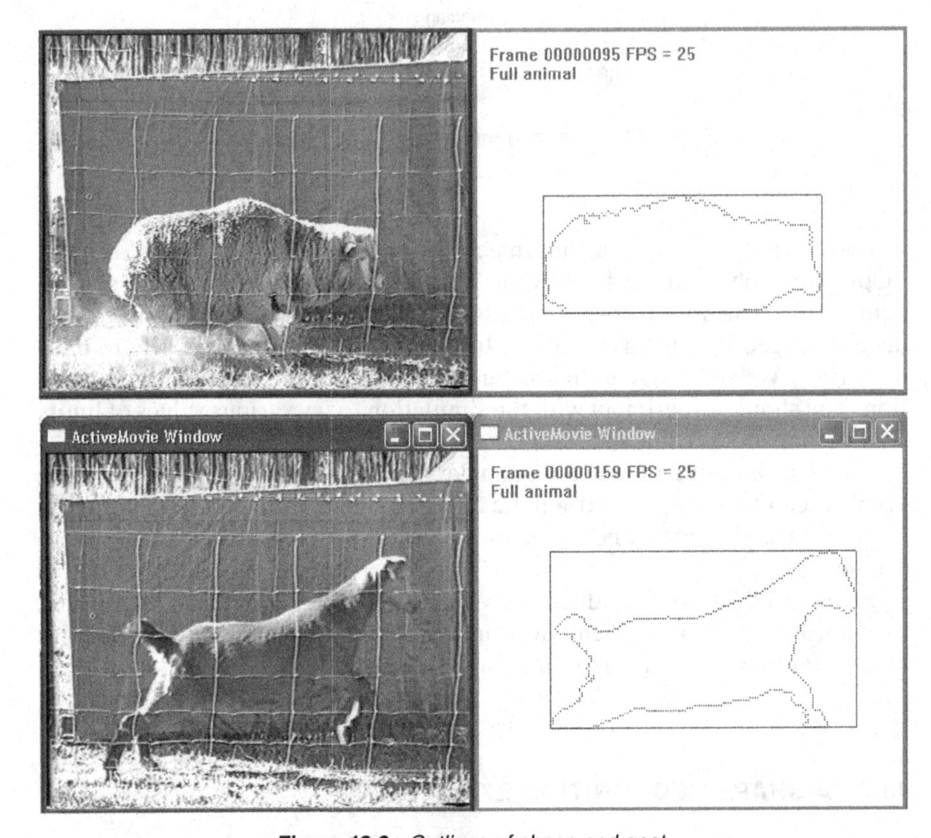

Figure 13.6 *Outlines of sheep and goat.*

approaching. An alternative is to look for changes in the background image, but this can give problems with animals moving in the background. Finding a workable compromise is in the nature of research. Image comparisons using sheep and goats are presented in Figures 13.6 and 13.7.

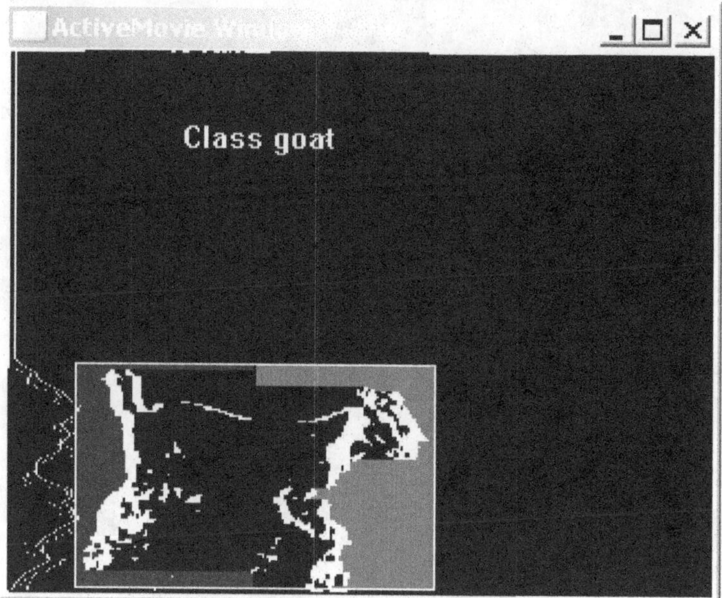

Figure 13.7 *Classification with natural background.*

14

The Human Element

14.1 THE USER INTERFACE

It is no use putting a heap of clever features into a device if the user is not comfortable with it. It is said that only a very few percent of videotape machines are ever programmed for automatic recording. So, why have the sales been so successful? Perhaps it is because a rental tape can be inserted into the slot and will start to play with no further instructions.

DVD players now perform the playing of rental videos—and they do not need to be rewound. It will be interesting to see how successful the hard-drive-based videorecorders will be in the long term. It will depend much on the simplicity of their programming.

14.1.1 What Do the Buttons Do?

Every one of the appliances that we take for granted was once a new product. There are museum-piece plaques that state; "This room is equipped with electric light." It must have been strange to look for the on/off control near the doorway, rather than a tap on the pipe of a gas lamp. But even today, all is not so simple. Which way do you move the switch to turn an electric light on?

An American will immediately say "up," but in Britain, Australia, and many other countries, the answer is "down."

It is obvious that each new product should try to build on the conventions that are already established in the mind of the user, but that does not always happen. Which way do you turn a faucet to increase the flow of water? Which way do you turn the knob of a radio to increase the volume?

"Which way" problems persist to this day. On a satellite or digital television receiver, it is not surprising that pressing the UP button on the remote will increase the channel number. But when the channel list is displayed on the screen, it is the DOWN button that increases the channel number because of our habit of writing larger numbers below smaller ones.

A succession of projects long ago on designing early digital controls for domestic cookers showed that the task of establishing the conventions was at least as challenging as writing the software. The operation had to be made totally intuitive. Nobody reads the manual in an appliance showroom, and that is where the purchasing decision is made.

The design task was actually even more complicated. The contract design work was being performed for the manufacturers of the electronic control box. They had to convince the cooker manufacturer that theirs was the controller to install in the new cooker range. The cooker manufacturer had to convince the stores to stock their brands of cooker and only then did the public get to see and try the new algorithms.

Let us start with the most fundamental function. How do you adjust the time? Even today the adjustment buttons of many clocks are designated "slow" and "fast." You hold the FAST button and watch the minutes and hours rip by. You release it some tens of minutes before the time you actually want. Then you press the SLOW button to let the minutes plod to the target.

But if impatience gets the better of you and the FAST button overshoots the target, you have to navigate another 24 h of adjustment.

So we designated the buttons "up" and "down." Clearly we needed to start the adjustment slowly and then speed up. But according to what algorithm? After numerous tests using the factory employees as guinea pigs, we settled on

(minute)—pause—(minute)(minute)(minute) until the hour is reached,
then (hour)—pause—(hour)(hour)(hour) . . .

Now we have to consider setting the cooking functions. Suppose that the current time is noon. You set the "ready time" for one o'clock. Then you enter "2 h" for the time that the meal should cook. When will the meal be ready? On some rival controllers, the answer would be "1:00 tomorrow." This was known as the "day factor error." Our simple remedy was patented and earned royalties and fees for infringement.

Increasing the "cook time" pushed the "ready time" correspondingly into the future, so that in the example above, a 2-h cooking time would give a 2 p.m. ready time. Selecting and advancing the "ready time" left the "cook time" unchanged, but wound up the "waiting time" before cooking would start. Decreasing the "ready time" reduced the waiting time, but when this reached zero, any further reduction was blocked.

To set the "ready time" for 1 p.m. tomorrow, it was necessary to wind it forward by all 23 h, something hard to do by accident.

14.1.2 What Sort of Display?

Long ago, the seven-segment display looked fashionable, whether in shadowy LCD (liquid crystal display), red LED, or flashing vacuum fluorescent figures. Today a mobile telephone is not complete without a glowing display screen that can show a full-color photograph. As the prices converge, the simplicity of numeric displays will continue to lose its advantage.

A seven-segment display (actually eight, with the decimal point) is easy to drive from the simplest of microprocessors, especially those designed to give the appropriate output levels. In fact, only one digit is illuminated at any one time. One "digit driver" line selects the digit, either pulling low a set of cathodes of an LED display or pulling high the digit's anode if the display is vacuum fluorescent. Meanwhile eight segment drivers cause the appropriate segments to be illuminated.

The display is kept refreshed by a background routine that lights each digit in turn. From this the processor is diverted to attend to any input or control actions.

Any more sophisticated display is likely to have a controller dedicated to its needs. The system designer's task is then to send it the data to display in the appropriate form over the appropriate bus connection.

Another form of output might not seem to fit the term "display," but it is a valuable user interface. It is sound output. Computers use sound to interact in a way that is often much more effective than vision. There is the responsive "click" of an input key, through ring tones and the annoying "ping" that alerts you to an error, all the way to voiced instructions telling you to "Please hold, your call is valuable to us."

A display has three functions. The first is to "close the loop" between user and computer, so that the user is assured that the programmed function is exactly what is wanted, whether it is a dialed telephone number or a Sunday roast. The second is to convey information to the user, such as a caller's identity or the speed of a jogger. The third is to look attractive at the point of sale, something dear to the heart of the client.

14.1.3 What Sort of Input?

The concept of a keyboard is firmly embedded in the folklore of computing, whether the teletypes of the ancient machines or the two or three buttons on an everyday appliance. A very few devices, such as intelligent pacemakers, might be designed without user inputs but most have an interface of some sort or another.

Pushbuttons present no real problem except that of laying them out in a way that will make their purpose intuitively obvious to the user. Should they be placed on the appliance itself, or should they be located on a remote control "zapper"? How should they be labeled? The international market decrees that the user must be preeducated with a set of basic concepts. A solid square means "stop," a triangle means "play," and two lines mean "pause." But what do you do if your product is truly novel?

The display can come to the aid of keys by displaying a descriptive legend against them, as in a cashpoint machine. It can even display the legend inside an image of the keys, if they have been replaced by a touchscreen. Now the user can be led through a complicated menu—sometimes rather reluctantly—where the functions of the keys change with each press.

There are other forms of input besides keys. The most familiar device that can input a nonnumeric quantitative input is the computer mouse. Drag it across a screen icon, and you can set the playback volume or the screen contrast, with no thought of entering numbers. The mouse has some very special features.

On a PC, as the mouse is moved, the cursor moves with it. When the cursor is over the feature or value the user wants to select, tapping a key or a touchpad will execute the desired action. But there is no absolute relationship between the cursor and the location of the mouse on the surface of the desk, or of the finger "tickling" the touchpad that substitutes for the mouse on many laptops. They simply cause the cursor to move. It is the computer that tells the user what action will be performed if the selection is clicked, so that there can be no disagreement.

On the other hand, a graphics tablet reads coordinates from the pen that the user is holding and a calibration error can cause the wrong action to be performed.

Touchscreens are useful if they have big, chunky legends but have serious problem with fine detail. By definition, the user's finger is between the screen and the eye, so that even if a spot on the screen lights to show the measured location of the finger, it is likely to be hidden!

The joystick is another popular input, both for games and for the setup movement of numerically controlled machine tools. It has even appeared as a substitute for a mouse, in the form of a small blob in the middle of the keyboard of some laptop computers.

Sound should not be forgotten as an input medium. Phonebook enquiries now let a computer try to understand the speech of the user, although in many cases a human operator must be called to the rescue. With the price of memory and processor power becoming vanishingly small, voice will soon be an attractive option for many pocket gadgets.

Vision is also a star that is likely to rise. Already a picture of a keyboard can be projected onto a flat surface, where the view of the user's fingers tapping away is translated into keystrokes. Gestures can be recognized and who knows, voice input might soon be made more reliable by lip reading.

14.2 IF ALL ELSE FAILS, READ THE INSTRUCTIONS

My wife proudly unpacked her new digital camera. It had come with the special offer of a 256-Mbyte memory card. With the card installed, she switched it on and took her first photograph. Nothing happened, except that the word FORMAT appeared on the miniature monitor screen.

In the slim handbook, we found a statement that the memory must be formatted before use. For details, see the full manual on the enclosed CD. Eventually, long after the photographic subject had gone, the procedure for formatting the memory was found on page 107 of a 10-Mbyte PDF file.

14.2.1 Designing the Handbook

The handbook is no less a part of the user interface than any software routine, although if the interface has been designed professionally, it should never need to be consulted. Problems can arise from too much as too little information. The help files of a well-known operating system are a good case in point.

In the early days of computing, help files were written by the same enthusiasts who had written the software. They might have lacked subtlety and polish, but they answered the needs of the user to the best of the writer's abilities. Then as the industry formed into large corporations, it was clearly felt that the writing of help files was a waste of programmers' time. It was easy to imagine an army of stenographers filling in boxes of a "What can we find to tell them about this?" questionnaire with no real thought for the needs of the user.

There has to be a fine balance between telling the user how to go about changing a setting, why they should want to change it, and what sort of values they would want to change it to. The tip "By selecting COMPRESSION in the dialog box, the value can be changed" does nothing to help the blood pressure.

Nowadays the help system seems most concerned with setting up the playing of video clips, changing the screen saver or choosing a pleasant color scheme. There seems to be little concern with the "real nitty gritty." Relentlessly searching for a technical term will leave the user adrift in an ocean of Web files. Alternatively they have to resort to "developer network" help disks, consisting of two or more CDs crammed with so many tips and hints that to find anything is like looking for a needle in a haystack.

But think again, the user has changed. Who is buying the most computers? For every engineer trying to do something innovative, there are a hundred would-be authors of the great novel, accountants, secretaries, lawyers, and lonely hearts searching the Web. The geeks and nerds are a long way down the pecking order in the consideration of a company that has certainly been successful in making a dollar or two.

The danger with leaving the help file or handbook task to the engineer is that the vital step that will baffle the user is so instinctive to the engineer that it is just not considered. Imagine trying to open a door if you have never seen a doorknob. There is an old joke about a new lumberjack who has spent all day cutting down a single tree with a chainsaw. Then when the supervisor starts it for him, he asks, "What's that noise?"

Nothing can beat the close observation of new users who are handed the device and asked to put it to work. It is their first questions and their first reactions that count. If the product is well designed, they will become expert within minutes or mere seconds and will no longer be suitable guinea pigs for a second test.

14.3 IT JUST TAKES IMAGINATION

Hardly a day goes by without media tales of wonderful new devices. Refrigerators with plasma displays and barcode and tag scanners will order replacement food from the supermarket via the Internet. Electric blankets and air-conditioning can be turned on by telephone text messages. "Swallowable" robots take biopsy samples and transmit a video travelog of their journey through the gut.

Some bright ideas can be a huge success, while many others vanish without trace. To some extent marketing may be the reason, but eventually it all comes down to the human element. With the power of embedded computers, if you can imagine it, you can probably build it. But can you sell it?

There are evolutionary products where technology nudges along the answer to the user's need, step by step. "Personal music" once took the form of a "ghettoblaster" balanced bulkily on a shoulder. Then the "Walkman" brought relief to us all, reducing the blast of sound to a merely irritating "Tsk tikatika tsk tika tsk" from the headset of a neighbor in the subway. Tape was supplanted by the compact disk, which in its turn has fought a losing battle with semiconductor memory and MP3 data compression, in the form of the "iPod."

The incredible shrinking memory chip is taking over portable storage applications everywhere. In "thumb drives" it has made the floppy disk history and in the digital camera it has sent photographic film the way of the phonograph.

Sometimes a brand new "need" is discovered, such as the mobile telephone. Technology has lifted the capabilities and reduced the price to make it universally available, but huge money is being made from spinoff markets. However, did teenagers communicate before text messaging? How much is being spent on downloading ring tones and games? Yet other attempts at "technology push" such as Internet access from a mobile phone seem to have met with an uphill battle.

So, when you start to work on your "better mousetrap," perhaps with machine vision identification of the mouse, MEMS sensors and actuators to close the trap, plus a wireless message to a cellphone to tell you to collect the trophy, first consider the human aspects. How will you convince me that I should buy it? How will I learn to set it? Where in the handbook does it tell me what to use for bait?

... and finally

The last sentence had been typed, the last figure drawn, but the task was far from finished. The text has been subject to the scrupulous attentions of a meticulous copy editor and must be marked up for the final edit.

Squeezing the text onto a narrower printed page has meant that many lines of computer code have "word-wrapped". I hope that I have caught them all, but if your computer grumbles about code that you have typed in from the text, the fragments of "left over" lines caused by word-wrap may well be the problem. You should be able to download an undamaged version from the book's web page at http://www.EssMech.com.

If you find that any of the promised material is missing from the website, please drop me an email at john@essmech.com—put "Essentials of Mechatronics" in the subject line so that the spam filter does not trap it! I will make room on the website for interesting questions and suggestions, too.

I hope that this book has convinced you of one important fact. Mechatronics can be fun.

Index

ADC (analog-to-digital converter), 33, 44
amplifier
  buffer, 102
  differential, 103
  non-inverting, 102
  summing, 100
  virtual earth, 99
angle from sensors, 241
assembly code, 25, 45, 93, 207, 208, 218
autopilot, 1, 6, 162, 186, 191

Ball and beam, 56
bicycle control example, 189

characteristic equation, 40, 185, 190
characteristic polynomial, 185
circuit theory, 95
code
  FUNCTION, 48
  structure, 51
  structured, 209
  SUB, 48
codec, 225
computer, 25
  embedded, 26, 204
  languages, 207
  memory, 25, 204
  processor, 25
  stack pointer, 205
computer language, 45
constraints, 196
control
  bang-bang, 200
  discrete time, 77
  dynamics, 39
  feedback, 36
  gain, 37
  minimum time, 200
  nested loops, 58, 191
  nonlinear, 58
  optimal, 199
  PID, 36
  pragmatic, 41
  sliding, 200
  state, 39
  theory, 36, 39
  topology, 185
  transfer function, 39

Essentials of Mechatronics, by John Billingsley

convolution, 157
coordinates
  Cartesian, cylindrical, spherical, 167
correlation, 159
cost function, 199

Darlington, 52
data compression, 225
Denavit-Hartenberg, 6, 172
D-H parameters, 177
difference equations, 150
Direct memory access (DMA), 206
DirectX, 27, 223
discrete time, 150
disturbances, 187
dynamics, 42, 179

EAROM, 211
edge tracing, 232
eigenvalue, 141, 171
eigenvector, 140, 171
embedded microcomputer, 43
embedded processor, 210
estimating velocity, 75
experiment, 43

feedback
  nested loops, 187
  non-linear, 187
feedback topology, 185
FET, 30
filter
  low-pass, 231
  switched, 105
Finite Impulse Response, 231
four terminal network, 98
frames, 173
Freeman Chain, 235
fuzzy logic, 198

gear, 161
gearbox, 41
GPS, 159
graphics, 46
Gray code, 19

H-bridge, 30, 70
hexadecimal, 50
hydraulic, 16

image
  binary, 226
  filter, 229
  grayscale, 226
integrator, 100
interrupt, 205
  timer, 242
inverse kinematics, 42, 178
inverted pendulum, 80
isocline, 194

Jacobian, 180
joystick, 243

key frame, 226
kinematic chain, 173
kinematics, 41

Laplace transform, 146
logic circuit, 113
low-pass filter, 48
lunar probe, 200
LVDT, 19

machine vision, 221
mark-space, 32
matrix, 131
  inverse, 138
  transpose, 136
  unit, 136
maximum principle, 200
mechanism, 161
microcomputer ADC, 211
Micromouse, 3, 6, 7
microprocessor, 211
motor, 161
  AC, 12
  axial field, 15
  back-e.m.f, 163
  brushless, 10
  DC, 10
  differential equation, 164
  drag cup, 14
  effective pulley, 164
  gear ratio, 164
  hysteresis, 15
  induction, 13, 31
  linear, 16
  permanent magnet, 162

single phase, 13
stepper, 10, 11, 15, 27, 50, 52
universal, 12

NEXT operator, 154
null string, 54

OCX, 223
operational amplifier, 35, 99
opto isolator, 32

partial differentiation, 179
phase plane, 192
phase-advance, 105
PIC, 211
pixel, 222
pneumatic, 16
polling loop, 205
position controller, 40
PRBS, 159
printer port, 50, 55
Printer port, 52
prismatic, 173
processor
embedded, 209, 210
proportional band, 41, 72
pseudocode, 209
pseudo-random binary sequence, 159

QBasic, 45

real-time, 47
regression, 247
Robocow, 237
robot, 41
mobile, 49, 52
revolute, 166
rotation, 169

sensor, 16
acceleration, 21
crossed Hall effect, 19
frame-scan, 222
Hall effect, 19, 82, 241
incremental encoder, 17
linescan, 23, 222
LVDT, 105
odometry, 238
phototransistor, 22
position, 16
potentiometer, 17, 19, 56
rate-gyroscope, 21
Sick, 222
tachometer, 21
velocity, 21
video, 24
vision, 22, 221
servomotor
AC, 13
shape recognition, 251
signal conditioning, 35
simple harmonic motion, 148
simulation, 120
analog, 128
position control, 125
singularity, 178
solenoid, 7, 10, 250
solid state relay, 31
s-psi curve, 235
stability, 40, 145, 149
bounded, 193
state equations, 143
discrete time, 152
matrix, 127
state variable, 40, 91, 98, 117, 118, 123, 150, 185, 189
steering, 249
subroutine, 204

tacho
DC motor, 21
tensor, 131
trajectory, 193
transfer function, 148
transformation, 136
transformation matrix, 170
transistor, 27
translation, 172
Turing machine, 203

Unimation Puma, 174

vector, 131
velocity control, 70
video, 27, 223
chip, 224
filter, 223

for Windows, 223
renderer, 224
vision guidance, 243
Visual Basic, 45

webcam, 24

z-transform, 154

Printed in the United States
By Bookmasters